An Introduction to Applied Electromagnetism

WILEY STUDENT SERIES IN ELECTRONIC AND ELECTRICAL ENGINEERING

Dr C. M. Snowden
Department of Electrical
and Electronic Engineering
Leeds University
UK

Dr A. McCowen
Department of Electrical
and Electronic Engineering
University College of Swansea
UK

Design and Technology of Integrated Circuits
D. de Cogan

An Introduction to Applied Electromagnetism
C. Christopoulos

An Introduction to Applied Electromagnetism

Christos Christopoulos
Reader in Electrical Engineering
University of Nottingham, UK

JOHN WILEY & SONS
Chichester · New York · Brisbane · Toronto · Singapore

Reprinted January 1998

Other Wiley Editorial Offices

John Wiley & Sons, Inc., 605 Third Avenue,
New York, NY 10158-0012, USA

Jacaranda Wiley Ltd, G.P.O. Box 859, Brisbane,
Queensland 4001, Australia

John Wiley & Sons (Canada) Ltd, 22 Worcester Road,
Rexdale, Ontario M9W 1L1, Canada

John Wiley & Sons (SEA) Pte Ltd, 37 Jalan Pemimpin 05-04,
Block B, Union Industrial Building, Singapore 2057

Library of Congress Cataloging-in-Publication Data:

Christopoulos, Christos.
An introduction to applied electromagnetism / Christos Christopoulos.
p. cm. — (Wiley student series in electronic and electrical engineering)
Includes bibliographical references.
ISBN 0 471 92761 9
1. Electric engineering. 2. Electromagnetism. I. Title. II. Series.
TK153.C455 1990 90-31983
621.3—dc20 CIP

British Library Cataloguing in Publication Data:

Christopoulos, Christos
An introduction to applied electromagnetism.
1. Electromagnetism. Applications
I. Title
537
ISBN 0 471 92761 9

Typeset by Thomson Press (India) Ltd., New Delhi.
Printed in Great Britain by Antony Rowe Ltd, Chippenham, Wiltshire

Πατέρα Δία, μα εσύ για βόηθα μαϛ απ'την πυκνήν αντάρα των Αχαιών τουϛ γιουϛ, ξαστέρωσε, δώσε να ιδούν τα μάτια, και μεϛ στο φωϛ πια τότε χάλα μαϛ, αφού το θέλειϛ τόσο!

'Ομηρόϛ Ρ 645

Father Zeus, break the gloom in front of the Achaeans, bring brightness, let our eyes see, and since you wish it destroy us but in the bright sunlight!

Homer, book 17/645

To Ranya, Kostas and Jason

Contents

Series Preface

The Wiley *EEE* textbook series in Electronic and Electrical Engineering has been developed to fulfil the requirements for course texts arising in universities and polytechnics. The series is aimed at the undergraduate level. The significant changes which have taken place in courses, stimulated by the Engineering Council initiatives and the IEE Accreditation system, have led to the need for texts which more closely match these new course structures. The Wiley *EEE* series takes a fresh look at the subject and aims to provide the reader with interesting and easily absorbed material.

An extensive survey of courses and opinions of staff in the many institutions within the UK was carried out prior to initiating the series. Every effort has been made to tailor the individual texts in the series to course requirements, whilst bearing in mind the need to produce texts which are attractive to students. It was felt particularly important to develop a style which included many examples, useful hints and a clear and easily followed layout.

Christopher M. Snowden, *University of Leeds*
Andrew McCowen, *University College Swansea*
October 1989

Preface

An Introduction to Applied Electromagnetism is a text suitable for students of Electrical and Electronic Engineering, Physics, Electronic Materials and other engineering and science-based courses. Emphasis is placed on the grasp of physical processes, the development of models for these processes, and their use in the study of engineering problems. The use of mathematics is minimized whenever possible. However, mathematical techniques are gradually introduced, especially in the second half of the book, in such a way that they will be a useful tool for the engineer and applied scientist. Part I covers basic electrostatics and magnetostatics and many useful applications. The mathematics required for this part is within the grasp of first-year engineering and science students. Parts II and III cover quasistatic and high-frequency phenomena. Familiarity with vector calculus should be achieved whilst working through this material.

I believe that the difficulties encountered by students of electromagnetism are not simply due to inadequate mathematical background. Indeed, students who have covered all the essential material in a mathematics course still find it difficult to relate the mathematical formalism to physical ideas and practical systems. An attempt has therefore been made in the text to render the connection between the mathematical model and the practical system plausible and easier to grasp. Several examples are given, followed by problems with answers. I have resisted the temptation to include a large number of problems. Most students are unlikely to tackle more than thirty problems as part of a 20-hour lecture course. I do regard however the problems included in the text as essential study material. Part I could be covered in a 20-hour lecture course and the remainder in another 20 hours.

In an introductory text several topics have to be omitted or only sketchily covered. I have tried to include topics which I regard as essential for all practitioners of electricity irrespective of specialization. In attempting to keep the treatment simple, some loss of rigour was unavoidable. The teacher can enhance some of the material, depending on the background and specialization of the class. There are many excellent texts with advanced material on electromagnetism. Some are listed in Appendix D.

Finally, I wish to thank my family and friends who, I am told, tolerated me during the antisocial periods when I was writing this book

C. Christopoulos, *University of Nottingham*
November 1989

CHAPTER 1

Introduction

Applied electromagnetism is concerned with the study of interactions between stationary and moving charges. It forms the basic intellectual framework on which our understanding of things electrical is based. Many engineering applications are founded on the understanding of electricity. Therefore both for intellectual and practical reasons the study of electromagnetism should be central in the formation of the professional electrical engineer and applied physicist. Yet the subject is considered difficult to grasp and is unpopular amongst students. The point is often made that the whole panoply of mathematics normally associated with the development of electromagnetism is unnecessary. Alternatively, it is stated that reliance on modern computer packages is all that is required to gain an understanding of this subject. Both these statements contain elements of truth, but miss the central objective in the teaching of electromagnetism. Like any other physical theory, electromagnetic theory seeks to construct models out of individual events which makes it possible to study fundamental properties, make predictions, and design engineering systems. These models are invariably mathematical in nature and in most cases they can be understood and used effectively without resort to complicated mathematics. As systems become more complex, the level of mathematical knowledge required increases. To handle this complexity numerical techniques and various computational tools must then be used. The aim of this textbook is to introduce the reader to the basic model-building process in electromagnetism, whilst at the same time to develop gradually the necessary analytical and numerical skills essential for more advanced work.

A recurring theme in model building is the identification of inherent limitations and assumptions. Emphasis is placed on this aspect throughout this book. Mathematical tools are introduced gradually as the need arises. An effort has been made to keep mathematics to a minimum. The reader, however, will benefit from trying to gain confidence with the mathematical manipulation of models which, after all, is the only way of gaining insight into complex interactions. A number of methodically worked examples are included together with problems. Answers to all problems are given. Material related to examples or problems is indicated with prefix E or P respectively. Appendix B contains material on vectors, differentiation and integration, and should be used whenever necessary. Appendix C contains information on physical constants and other data. References for further reading are given in Appendix D. The international system of SI units is used throughout the text. Solving the problems included in the text forms an essential part of the learning process and it is hoped that the reader will attempt most of them.

Part I of the text deals essentially with static situations (i.e. charge distributions which do not change with time). Part II deals with problems where slow variations in charge distribution take place, whilst Part III deals with phenomena associated with rapidly accelerating charges.

PART I

PHENOMENA ASSOCIATED WITH STATIC CHARGES OR WITH CHARGES MOVING AT CONSTANT VELOCITY

Electromagnetism deals with the interactions between electric charges. These charges may be constrained inside conductors, insulators or semiconductors, or, they may be free to move in vacuum or in ionized gaseous media (plasmas). A vast range of problems is thus encountered, in which various interaction mechanisms are thought to be dominant. For the student beginning to learn electromagnetism, the presentation of such complexity in its entirety must be bewildering. Initially, it is therefore sensible to limit the scope of the treatment to simple situations, where the basic concepts can be gradually developed, the mathematical models constructed and application areas illustrated. It would, however, be dangerous to imagine that what we do under these circumstances will generally apply to all the problems we are likely to encounter. Whilst making simplifications and assumptions to produce models which are easy to handle, we must not forget their limitations.

In the first part of the book, the underlying assumption is that charges are essentially stationary or, if they move, that they do so at a constant velocity. We thus make what is known as the static approximation. Occasionally, as when charging a capacitor or energizing an inductor, this is far too severe an assumption and then improvements to our model must be made. The major simplification stemming from the static approximation is that the interaction between static charges may be studied in terms of an influence described as the *electric field*, whilst interactions between moving charges (e.g. steady currents) can be studied in terms of an influence described as the *magnetic field*. The two fields, in this approximation, are independent of each other. The mathematical models thus developed are not very complex and a wide class of problems may be studied with the minimum of mathematical effort. The reader may well wonder how justifiable such an approximation is. An example perhaps should indicate that such an approach is reasonable: consider a two-wire line connected at one end to a battery. The other end is either an open circuit or a short circuit. In the former case very little current flows upon connecting the battery to the line. An electric field is established between the wires, the magnetic field being insignificant. In the latter case, however, a significant current flows and a magnetic field is established, the electric field being insignificant. The open-circuit problem is typical of an electric field system and is studied in Chapter 2. The short-circuit line is typical of a magnetic field system and is studied in Chapter 3.

CHAPTER 2

Electric Field Systems

These systems are typical of situations where the interacting electric charges are essentially static. The types of questions addressed here are: How can the force between charges be calculated? What is meant by potential difference? How can potential difference be calculated? How do different materials behave in the proximity of charges? How can a capacitor be built and its capacitance calculated? Are there limits to the energe that can be stored in a capacitor?

2.1 FORCES BETWEEN ELECTRIC CHARGES

It is a common experience that when an apple is released from the hand, it falls to the ground. To comprehend this observation humans have produced an elaborate story which runs more or less as follows. An attribute is assigned to matter called mass. Both the apple and the Earth thus have a mass and it is further postulated that masses attract each other. It is not then difficult by observation to establish a mathematical model suitable for the calculation of the force acting on any mass. It is found that this force, known as the gravitational force, is proportional to the product of the interacting masses and inversely proportional to the square of their separation. The constant of proportionality depends on the units used to measure force. A whole range of observations have established that the model just described makes predictions which are good enough and the model is thus generally accepted.

There are, however, forces which cannot be accounted for by the gravitational model just described. It appears that these forces do not depend on the mass. Instead, another attribute is assigned to matter called the *electric charge*, q, and models are constructed, similar to the gravitational model, in which now the charge q plays the leading role. Moreover it is found that there are two types of charge, namely negative and positive charge. The electron is the constituent of matter which is responsible for the negative charge. Charge is measured in coulombs (C) and symbols q or Q are used for its description. Measurements of force between localized (point) charges have led to a mathematical model for this interaction, embodied in Coulomb's law:

$$F = K_e \frac{|Q_1||Q_2|}{r^2} \tag{2.1}$$

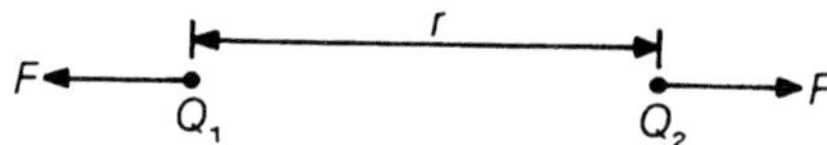

Figure 2.1 Forces on point charges.

where F is the force in newtons between the two interacting charges of magnitudes Q_1 and Q_2, measured in coulombs, and r is their separation in metres as shown in Figure 2.1.

Equation (2.1) gives the magnitude of the force. Its direction is as shown in Figure 2.1 (i.e. repulsive) for the case of two charges of the same polarity. The force is attractive if the charges are of opposite polarity. The constant K_e depends on the units used and on the surrounding medium. For SI units and for charges in vacuum (or to a good approximation in air) $K_e = 1/4\pi\varepsilon_0$, where $\varepsilon_0 = 8.85 \times 10^{-12}\,\mathrm{F\,m^{-1}}$ is the dielectric permittivity of vacuum (or air). Hence Coulomb's law in free space is:

$$F = \frac{1}{4\pi\varepsilon_0}\frac{|Q_1||Q_2|}{r^2}. \tag{2.2}$$

EXAMPLE E1

Two charges $Q_1 = 1.5\,\mu\mathrm{C}$ and $Q_2 = -2\,\mu\mathrm{C}$ are placed 8 cm apart. Calculate the magnitude and direction of the Coulomb force.

Solution

$$F = \frac{1}{4\pi \times 8.85 \times 10^{-12}}\frac{(1.5 \times 10^{-6})(2 \times 10^{-6})}{(0.08)^2} = 4.21\ \mathrm{N}.$$

Since the charges are of opposite polarity the force is attractive.

Consider now the system of three charges shown in Figure 2.2. The sign next to each charge indicates its polarity. Thus Q_1 $(-)$ means a negative charge of magnitude. Q_1. The magnitude and direction of the force acting on Q_3 $(+)$ is required. This problem can be tackled easily by applying the *principle of superposition*. Since the force F_{31} acting on Q_3 due to Q_1 is proportional to Q_1

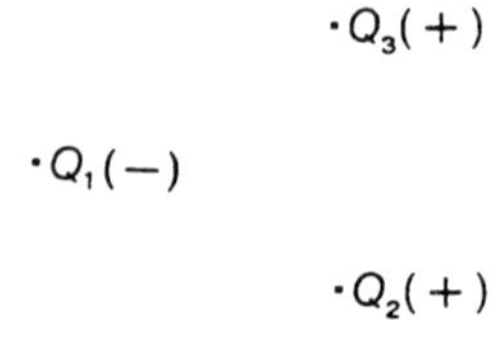

$\cdot Q_3(+)$

$\cdot Q_1(-)$

$\cdot Q_2(+)$

Figure 2.2

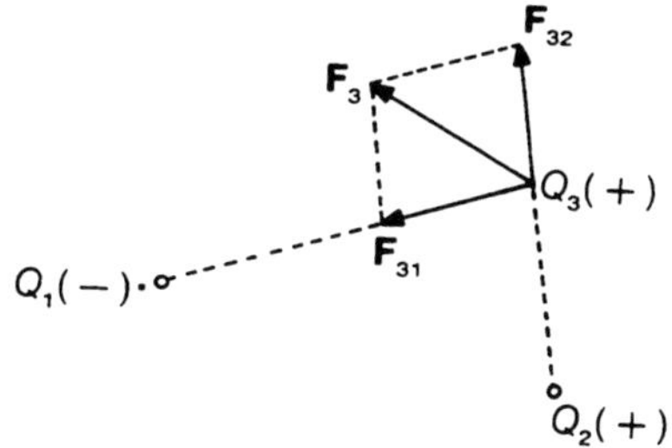

Figure 2.3 Force $\mathbf{F}_3$ acting on charge Q_3.

(i.e. there is a linear relationship between force and charge) and similarly for the force F_{32} due to Q_2, the total force F_3 on Q_3 due to Q_1 and Q_2 acting together is equal to the sum of F_{31} and F_{32}. But, this summation must take account of the directions of F_{31} and F_{32}. This is shown in Figure 2.3. Readers familiar with vector notation will realize that the summation required to find the total force is a *vector summation*:

$$\mathbf{F}_3 = \mathbf{F}_{31} + \mathbf{F}_{32}$$

REMARKS

It is a gross error to combine forces by simply adding their magnitudes. Except under special circumstances (e.g. forces all acting in the same direction) a vector summation is required.

Superposition may be used in systems which are linear, i.e. where the response is proportional to the stimulus or excitation of the system. In systems in which proportionality does not apply (non-linear systems) superpositiom cannot be used. An ordinary resistor is a linear device since when the voltage across it doubles, so does the current. However, repeatedly doubling, the voltage will overheat the resistor and proportionality between voltage and current will be lost. When this happens, the resistor behaves as a non-linear device. Most devices or systems are linear only within a specified range of parameters (in their linear region of operation). Some devices, such as diodes, are inherently non-linear during normal operation.

EXAMPLE E2

Find the magnitude and direction of the force acting on charge Q_3 in Figure E2.

Solution

Calculate the magnitude of the forces using Coulomb's law. The force acting on

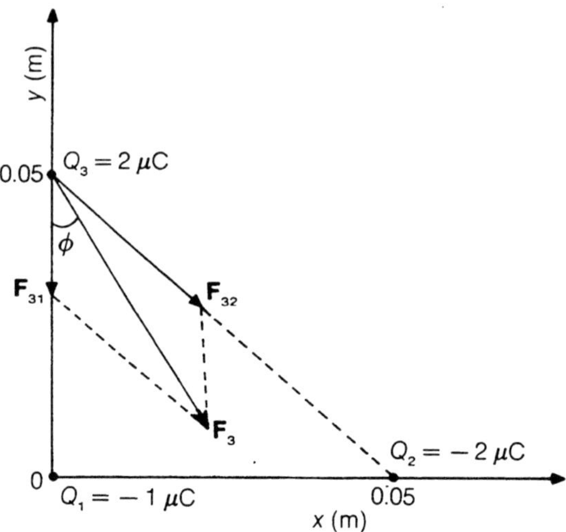

Figure E2 Force $\mathbf{F}_3$ on Q_3 due to Q_1 and Q_2.

Q_3 due to Q_1 is

$$F_{31} = \frac{1}{4\pi\varepsilon_0}\frac{|Q_1||Q_3|}{r^2} = \frac{1}{4\pi \times 8.85 \times 10^{-12}}\frac{(1 \times 10^{-6})(2 \times 10^{-6})}{0.05^2} = 7.19\ \mathrm{N}$$

and similarly the force on Q_3 due to Q_2 is

$$F_{32} = \frac{1}{4\pi \times 8.85 \times 10^{-12}}\frac{(2 \times 10^{-6})(2 \times 10^{-6})}{0.05^2 + 0.05^2} = 7.19\ \mathrm{N}.$$

Both forces are attractive as shown in Figure E2. The total force acting on Q_3 is found by superposition; $\mathbf{F}_3 = \mathbf{F}_{31} + \mathbf{F}_{32}$. This vector sum may be obtained by graphical means as shown in Figure E2. Alternatively, the components of $\mathbf{F}_3$ in the x- and y-directions may be found by adding the corresponding components of $\mathbf{F}_{31}$ and $\mathbf{F}_{32}$. F_3 in the x-direction $= F_{3x} = F_{31}\cos 90^\circ + F_{32}\cos 45^\circ = 5.08$ N. Similarly F_3 in the y-direction $= F_{3y} = -F_{31} - F_{32}\cos 45^\circ = (-7.19 - 5.08) = -12.27$ N. Hence $F_3 = (F_{3x}^2 + F_{3y}^2)^{1/2} = (5.08^2 + 12.27^2)^{1/2} = 13.28$ N and $\phi = \tan^{-1}(5.08/12.27) = 22.5^\circ$.

Let us consider again Coulomb's law (equation (2.2)) showing the force acting on a charge Q_2 due to a charge Q_1. Equation (2.2) may be written as follows:

$$F = \left(\frac{|Q_1|}{4\pi\varepsilon_0 r^2}\right)|Q_2|. \tag{2.3}$$

The modification of (2.2) shown in (2.3) is trivial from the mathematical point of view as it simply brackets together some of the terms. Equation (2.3) signals, however, another way of looking at the interaction between Q_1 and Q_2. The conceptual model of this interaction can be described as follows: Charge Q_1 sets up in its vicinity an influence, described as an *electric field*, E, such that any charge

Q_2 in the vicinity of Q_1 will experience a force $F = |Q_2|E$. For the particular example of point charges $E = |Q_1|/4\pi\varepsilon_0 r^2$, but for other charge distributions different formulae apply. For the example depicted in Figure 2.3, the force acting on Q_3 is $F = |Q_3|E$, where E is the electric field produced by Q_1 and Q_2 (i.e. the combined influence of the two charges) at the location of charge Q_3. For other charge distributions similar formulae apply although the calculation of the electric field may be quite involved.

In addition to the magnitude of the force, its direction is also required and this can be obtained if the electric field at each point is given a direction as well as magnitude (both force and electric field are thus vector quantities). The electric field is derived in such a way, that the forces acting on positive and negative charges are parallel and antiparallel to **E** respectively. Hence the Coulomb force acting on charge Q placed at a location where the electric field is **E** is

$$\mathbf{F} = Q\mathbf{E}. \tag{2.4}$$

Let us now examine what the electric field due to a positive charge Q looks like (Figure 2.4).

Choose a point P, which is in no way special, a distance r away from the charge Q. If a test charge Q_1 (+) is placed at P, it will experience a force which from Coulomb's law is $F = |Q||Q_1|/4\pi\varepsilon_0 r^2$ and is repulsive. Hence, if the electric field at P is chosen in the direction shown in Figure 2.4 and has magnitude $E = |Q|/4\pi\varepsilon_0 r^2$, the force **F** can also be obtained directly from the expression $\mathbf{F} = Q_1\mathbf{E}$. Moreover, if a negative charge is placed at P the same formula will apply provided the charge is entered with a negative sign.

It would be a confusing way of indicating the electric field due to Q, if the whole space around Q were filled with arrows of length $|Q|/4\pi\varepsilon_0 r^2$ and radial direction. Instead, the electric field is depicted as shown in Figure 2.5. The lines shown are

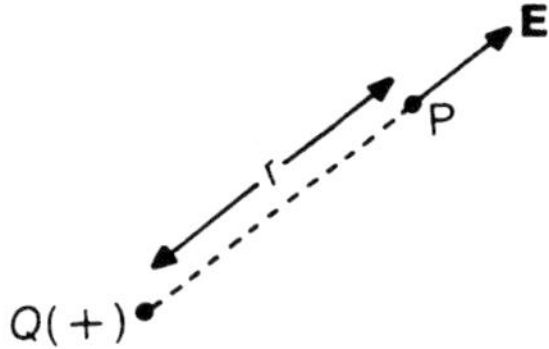

Figure 2.4 The electric field **E** at point P due to charge Q.

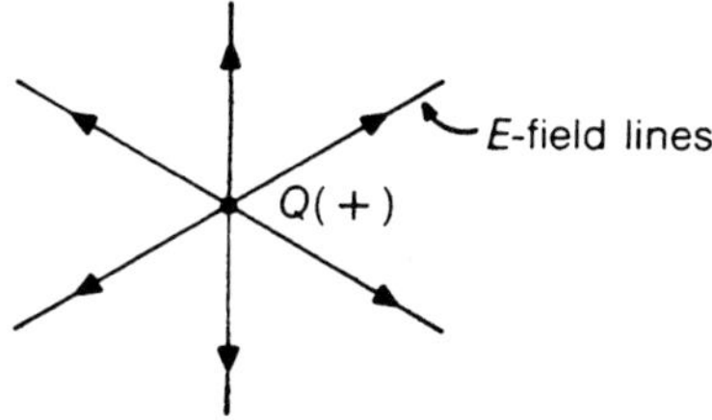

Figure 2.5 Electric field lines around a positive point charge.

called the E-field lines. They are radial since the field is everywhere radial, and they are symmetrically placed since there is no preferred direction for the field around Q. If a test charge were placed in space it would move along one of these radial lines. At first sight it may be thought that Figure 2.5 conveys no information about the magnitude of the electric field. This is not true—the lines are closer together where the field is strongest (i.e. nearest to the charge). If the charge in Figure 2.5 was negative, the E-field lines would be identical but the arrows would point towards the negative charge. More examples of visualizing the electric field will be given in section 2.6.

EXAMPLE E3

For the charge configuration in example E2, calculate the electric field at a point with coordinates (0, 0.05 m) due to charges Q_1 $(-)$ and Q_2 $(-)$. Hence calculate the force acting on Q_3 $(+)$ using equation (2.4).

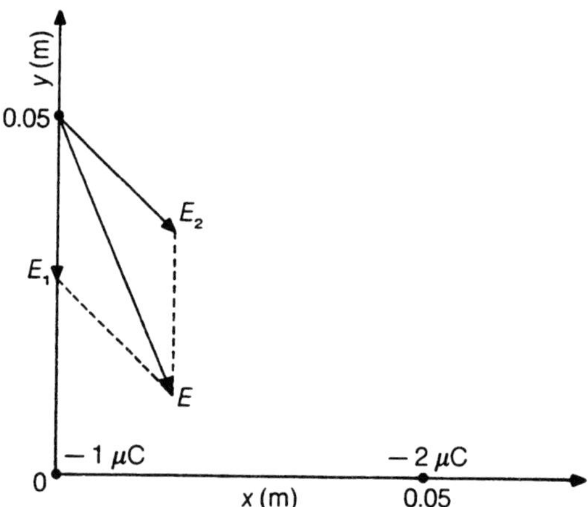

Figure E3 Electric field **E** at the position of charge Q_3.

Solution

The electric field at point (0, 0.05 m) due to Q_1 is

$$E_1 = \frac{|Q_1|}{4\pi\varepsilon_0 r^2} = \frac{10^{-6}}{4\pi\varepsilon_0 \times 0.05^2} = 3.6 \times 10^6\ \mathrm{V\,m^{-1}}.$$

Since Q_1 is negative, the field E_1 is directed as shown in Figure E3. Similarly, the field due to Q_2 is

$$E_2 = \frac{2 \times 10^{-6}}{4\pi\varepsilon_0 \times 0.707^2} = 3.6 \times 10^6\ \mathrm{V\,m^{-1}}.$$

The total field is found by superposition: $E = 2 \times 3.6 \times 10^6 \cos 22.5° = 6.65 \times 10^6$ V m^{-1}. Hence the force acting on Q_3 is $F_3 = Q_3 E = 2 \times 10^{-6} \times 6.65 \times 10^6 = 13.3$ N, as found in example E2.

REMARKS

There are two important things to remember about electric charges.

Firstly, ordinary objects are to a very high degree of accuracy electrically neutral. This does not mean that they do not contain electric charges, but that they contain equal amounts of positive and negative charge so that, overall, they behave as neutral bodies. In establishing electric fields all that is done is to disturb this equilibrium by separating charges in some way.

An isolated negative charge is a convenient mathematical abstraction. In reality, it is obtained by leaving some other body with a positive charge of equal magnitude. Figure 2.5 makes sense only if a charge Q ($-$) is distributed uniformly far away from Q ($+$) (at infinite distance).

Secondly, positive and negative charges cannot be destroyed or created from nothing—charges live for ever! In any practical situation the total charge must therefore remain constant. This statement is known as the principle of charge conservation and is a very useful tool in tackling practical problems.

SUMMARY

After studying this section, the reader should be familiar with the calculation of force between charges (Coulomb's law, equation (2.2)). The force on a charge Q may also be obtained from equation (2.4), where the electric field is used to represent the combined influence of all other charges on Q. The electric field may be viewed as the agent through which the combined influence of charges is exercized on Q. To define the force or the electric field, their amplitude and also their direction must be specified—both are vector quantities. The combination of forces acting on a charge, or of electric field components due to several charges, is always a vector addition.

Although the formulae given in this section are for fields due to point charges, they may be easily extended to deal with any charge distribution. This is achieved by splitting the actual charge distribution into parts small enough to be regarded as point charges. The field due to each point charge is then calculated and the total field found by superposition. However, this method of calculating electric fields is not particularly easy to use, except when a computer is available to do the calculation. A more elegant method for obtaining the electric field due to simple charge distributions is described in the next section.

PROBLEM

P1 Two point charges, $Q_1 = 2\,\mu\text{C}$ and $Q_2 = -3\,\mu\text{C}$ are held apart as shown in Figure P1. A third point charge $Q_3 = 1.5\,\mu\text{C}$ is placed at point A as shown.

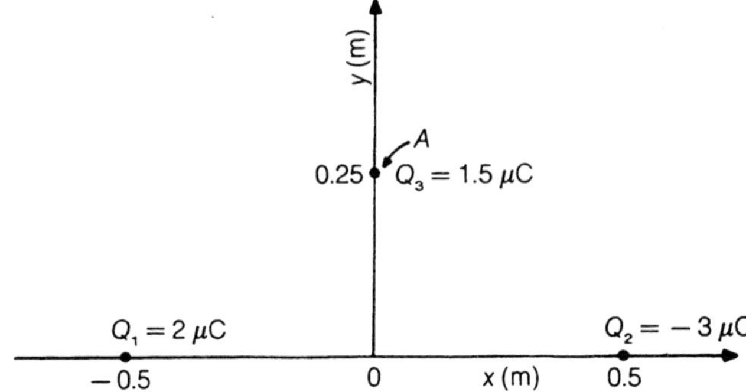

Figure P1

(*a*) Calculate the magnitude and direction of the force acting on Q_3 using Coulomb's law.
(*b*) Calculate the electric field at A due to charges Q_1 and Q_2. Confirm that the force calculated from $\mathbf{F} = Q\mathbf{E}$ is the same as found in (*a*).
(*Answer:* (*a*) $|F| = 0.194$ N, (*b*) $E_x = 12.87 \times 10^4\ \mathrm{V\,m^{-1}}$, $E_y = -1.29 \times 10^4\ \mathrm{V\,m^{-1}}$.)

2.2 THE RELATIONSHIP OF THE ELECTRIC FIELD TO ITS SOURCES—GAUSS'S LAW

In the last section, a procedure was established for calculating the electric field due to point charges. It was found that the field due to a charge Q at a point a distance r away from Q is $E = Q/4\pi\varepsilon_0 r^2$, and that it points radially away from a positive Q or radially towards a negative Q. If more than one point charges is present, then the electric field due to each point charge acting on its own is calculated, and the total field is found by combining all these partial field components (i.e. by invoking superposition).

In this section an alternative procedure, known as Gauss's law, is described, which permits the calculation of fields due to a known charge distribution. In many cases, Gauss's law is easy to apply, and it quantifies in an elegant way the manner in which charges establish around them an influence described as an electric field.

Consider a charge distributed inside a volume V, as shown in Figure 2.6. The total charge Q inside V may or may not be distributed uniformly. Visualize also a surface S, of arbitrary shape, which must, however, completely enclose the charge distributed in V. The surface S is then subdivided into small patches, each of area $\Delta S_1, \Delta S_2, \Delta S_3, \ldots$ until the entire surface S is covered by these patches. The electric field in the middle of the patch ΔS_1 is say $\mathbf{E}_1$, as yet unknown. It is not necessarily perpendicualr to patch S. The direction normal to ΔS_1 is shown by the line $\hat{n}$ and suppose that $\mathbf{E}_1$ makes an angle φ with $\hat{n}$. The component of $\mathbf{E}_1$ normal to ΔS_1 is then $E_{1n} = E_1 \cos\varphi$. Similar calculations may be done for all patches $\Delta S_1, \Delta S_2$, etc, and the following sum formed: $\pm \Delta S_1 E_{1n} \pm \Delta S_2 E_{2n} \pm \cdots$ until all patches are accounted for. The plus sign is chosen when the electric field is pointing out of the

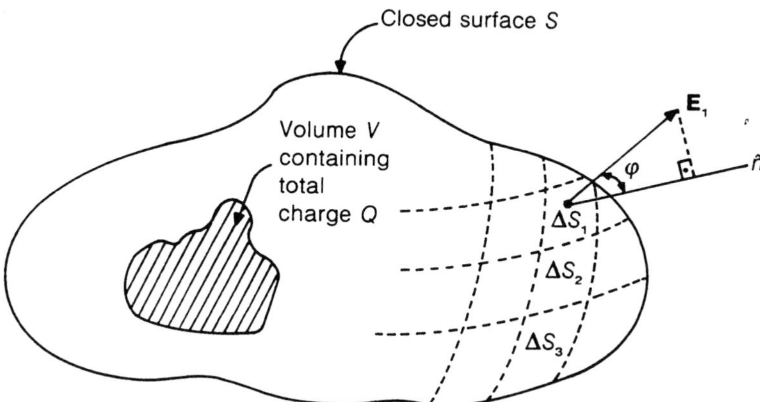

Figure 2.6 Gaussian surface S surrounding charge Q distributed in volume V.

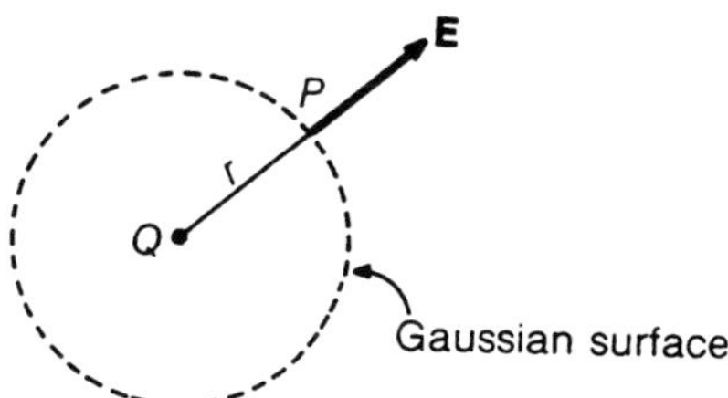

Figure 2.7 Gaussian surface suitable for the calculation of the electric field around a point charge Q.

surface, and the negative sign when it is pointing into the surface. Gauss's law expresses the remarkable fact, that the sum thus calculated is always equal to the charge Q enclosed by S divided by ε_0, i.e.

$$\pm \Delta S_1 E_{1n} \pm \Delta S_2 E_{2n} \pm \cdots = Q/\varepsilon_0 \tag{2.5}$$

or, in a more compact form,

$$\sum \pm \Delta S E_n = Q/\varepsilon_0 \tag{2.6}$$

where the summation is done over all the patches covering S and the appropriate sign is chosen as explained above.

The important fact is that, irrespective of the shape of surface S (as long as it completely encloses Q), irrespective of how the patches are chosen (as long as they are not too large), the sum in the left-hand side of (2.5) is equal to the total charge Q enclosed by S divided by ε_0. The charge Q may be distributed in any way inside S. Equation (2.5) is worth remembering, as it expresses in a very general way the relationship of the electric field (E) to its sources (Q).

Let us check that (2.5) gives the correct answer for the electric field at a point P a distance r away from a point charge Q (Figure 2.7). To apply Gauss's law a surface must be considered which completely encloses Q and, since the field at point P is required, passes through point P. Any surface will do but, in order to avoid getting lost in mathematics, it pays to choose *a* simple surface which will give the desired results with the minimum of effort. It is good practice to choose a surface which exploits the symmetry of the problem. For the charge configuration of Figure 2.7, a spherical surface centred at Q and passing from P offers several advantages. Firstly, everywhere on this surface the magnitude of **E** is the same, since owing to symmetry there can be no preferred direction for the influence set up by Q. Secondly, the field everywhere around Q must be radial, since if it were not so there would be a preferred direction, thus violating the symmetry of the problem. Hence, **E** is everywhere normal to the surface chosen. Applying Gauss's law

$$\sum \pm \Delta S E_{\mathrm{n}} = Q/\varepsilon_0.$$

But the electric field on each patch ΔS on surface S is normal to S, i.e. $E_{\mathrm{n}} = E(r)$ the same for all patches ΔS but, possibly, depending on r. Hence, $E(r)$ can be factored out of the sum giving $E(r) \sum \Delta S = Q/\varepsilon_0$. The sum $\sum \Delta S$ is simply the total surface area S which, in this example, is equal to $4\pi r^2$. Hence, as expected, $E = Q/4\pi\varepsilon_0 r^2$.

This derivation is not a proof of Gauss's law; it simply shows that it works for a simple case. A proof of Gauss's law is given in Appendix A.

Equation (2.6) may be multiplied by ε_0 to obtain $\sum \pm \Delta S(\varepsilon_0 E_{\mathrm{n}}) = Q$. The quantity $\varepsilon_0 E_{\mathrm{n}}$ is so important as to warrant its own name. It is called the normal component of the electric flux density $D_{\mathrm{n}} = \varepsilon_0 E_{\mathrm{n}}$ and in a sense it projects the influene of the charge Q at a distant surface element ΔS. By analogy, the *electric flux density* in free space is defined by the expression:

$$\mathbf{D} = \varepsilon_0 \mathbf{E} \tag{2.7}$$

where **D** is a vector of magnitude $\varepsilon_0 E$, having the same direction as **E**. It has dimensions of charge per unit area and it is measured in $\mathrm{C\,m^{-2}}$. If hypothetical charges of density equal to D_{n} were distributed on surface S, and the original charge Q was removed, the influence of Q outside S would remain unchanged. Hence **D**, like **E**, describes in some way the influence of charges at points distant from them. In terms of the electric flux density, Gauss's law is:

$$\sum \pm \Delta S D_{\mathrm{n}} = Q. \tag{2.8}$$

Gauss's law is an extremely powerful tool in the study of electric field problems. Its practical use, however, is limited to situations where the field possesses some symmetry. In such cases, by a suitable choice of the surface S, it is possible to obtain useful results with little mathematical effort. The choice of this surface S, which hereafter will be called the Gaussian surface, requires skill and some understanding of the field. Several examples are given in section 2.6, after Gauss's law has been adapted to deal with fields in materials other than air. A simple example, however, will be given now to illustrate the application of the law.

EXAMPLE E4

A positive charge is distributed uniformly along a long line, so that there are q_l Coulomb per metre of line length. Calculate the electric field due to this line charge.

Solution

This is a classic example of applying Gauss's law to obtain the electric field. The alternative would be to divide the line into short segments and then find the field due to each segment assuming that it behaves like a point charge. The fields due to each segment must then be combined (as vectors) to find the total field. This procedure, although possible, is quite a lengthy one. Let us stick with Gauss's law.

The first task before applying Gauss's law is to select a Gaussian surface. Any closed surface will do, but it pays to choose one that exploits the symmetry of the problem. It is therefore essential, before embarking into the mathematics, to ascertain, using physical intuition, whether anything could be done to facilitate the mathematical formulation of the problem. For such a simple distribution it is not

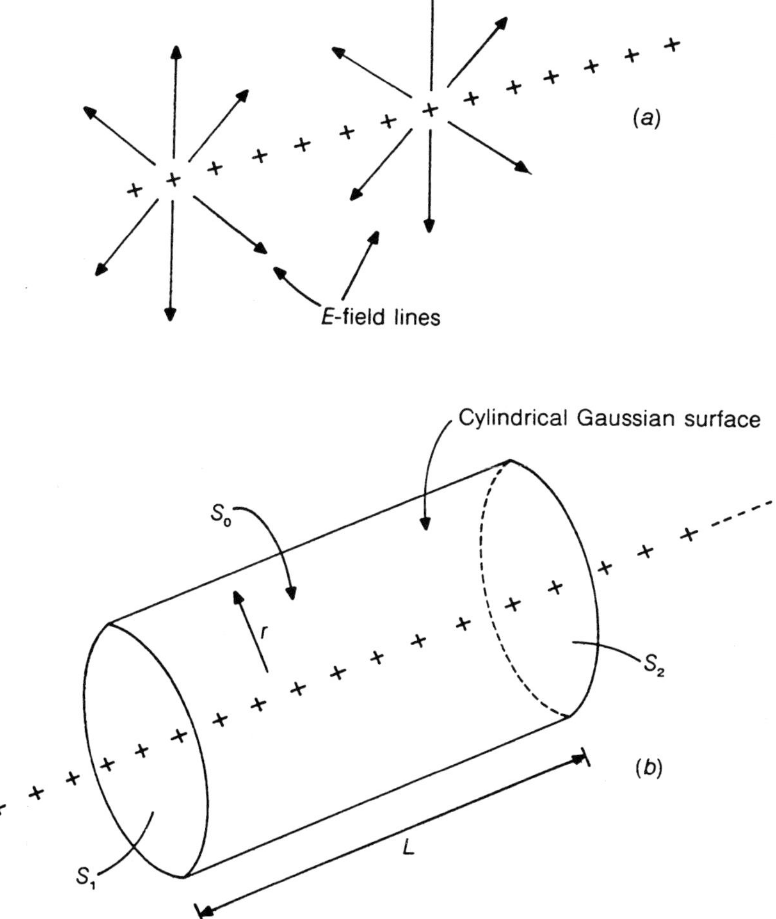

Figure E4 Determination of the electric field around a positive line charge.

difficult to visualize the electric field. This is shown in Figure E4(*a*). The following statements can be made about the electric field:

(i) it does not depend on the position along the line charge (since the charge has infinite length);

(ii) as long as the field is measured at points the same distance r away from the line charge, it will have the same magnitude (since there can be no preferred direction around the line charge);

iii) on a cylindrical surface coaxial with the line charge, the electric field will have constant magnitude and radial direction.

The electric field has therefore cylindrical symmetry, and a cylinder coaxial with the charge is chosen as the Gaussian surface, as shown in Figure E4(*b*). The cylinder is arbitrarily chosen to have a length L and radius r. The surface of this cylinder will be the Gaussian surface. It must be a closed surface, hence it consists not only of the curved part S_0, but also of the flat ends S_1 and S_2. Gauss's law states that:

$$\sum \pm \Delta S \mathrm{D_n} = \text{total charge enclosed},$$

where, the summation is done over all patches ΔS covering S_0, S_1, S_2.

Let us start with the two flat sides S_1 and S_2. These must be divided into small patches and the area of each patch multiplied by the component of the electric flux density D_n normal to the surface. However, from the symmetry of the problem, the electric field and hence $\mathbf{D}$ is everywhere radial and therefore parallel to S_1 and S_2. D_n is hence equal to zero on surfaces S_1, S_2. The only contribution must come from surface S_0. Here the field is radial, hence normal to S_0. In addition, it has a magnitude $D(r)$, constant at all points on S_0. Hence, Gauss's law is

$$\sum_{\text{over } S_0} \Delta S\, D(r) = \text{total charge enclosed.}$$

Since $D(r)$ is constant over S_0, it can be factored out to give

$$D(r) \sum_{\text{over } S_0} \Delta S = \text{total charge enclosed.}$$

The sum over all patches in S_0 is simply the area of the curved surface $2\pi rL$ and the total charge enclosed is the charge per unit length times the length of the cylinder $q_l L$. Hence, $D(r)2\pi rL = q_l L$, giving for the flux density $D(r) = q_l/2\pi r$ and for the electric field $E(r) = q_l/2\pi\varepsilon_0 r$.

The choice of this particular Gaussian surface has made the calculation very simple. Before completing this example, it is useful to check the units in the formula for $D(r)$. Electric flux density is measured in C m^{-2}, hence the expression $q_l/2\pi r$ must also result in the same units. The formula checks out since q_l is in C m^{-1} and r in m. This simple check is highly recommended, since in case of inconsistencies, it alerts one to the possibility of calculation errors.

REMARKS

Gauss's law, as expressed by equation (2.8), was formulated in a way that avoids the use of vectors. Readers familiar with vector notation and calculus will benefit from using these more powerful mathematical tools. To do this a vector **ds** is defined on each patch, which is vertical to and points outwards from the surface S. Its magnitude is simply the surface area of the an infinitesimally small patch on S. Then summation over all patches on S, becomes an integration over S and $\Delta S\, D_n$ becomes $\mathbf{D}\cdot\mathbf{ds}$. The latter expression is called the dot product of **D** and **ds** and it is equal to $D\,\mathrm{d}s \cos\varphi$, where φ is the angle between the directions of D and ds. Since ds is normal to S, then $D \cos\varphi$ is the component of D normal to S, i.e. D_n. Hence Gauss's law may be expressed in vector form as

$$\int_S \mathbf{D}\cdot\mathbf{ds} = \text{total charge enclosed.} \tag{2.9}$$

The integral is evaluated over the Gaussian surface S and is an example of a surface integral. It is no use remembering (2.9) if the reasoning and calculation procedure leading up to equation (2.8) is not fully understood.

In this section one of the fundamental laws of electromagnetism, Gauss's law, has been presented, relating the electric field to its sources. The law is applied on a closed surface (Gaussian surface), which may be chosen arbitrarily, but, which in practice is chosen in a manner that facilitates calculation. The normal component of the flux density, multiplied by the area of small patches on the Gaussian surface is evaluated as indicated by equations (2.8) or (2.9). Gauss's law states that the sum of all such terms, covering the entire surface, is equal to the total charge enclosed by the surface.

When applying Gauss's law the following rules must be observed:

(i) For case of calculation, the Gaussian surface must be chosen in a manner that exploits the symmetry of the problem.

(ii) A preliminary assessment must be made of the properties of the field, e.g. identification of areas where the field is constant.

(iii) Electric field components emerging from the surface contribute positive terms to the Gaussian integral, whilst field components directed into the surface contribute negative terms.

(iv) The only charge that needs to be taken into account in (2.8) or (2.9) is the charge enclosed by the Gaussian surface.

Up to this point, two methods for calculating fields have been introduced. Firstly, fields may be calculated by dividing charge distributions into small elements, and then using formulae for point charges and superposition to obtain the total field. Secondly, fields may be obtained by applying Gauss's law.

Whenever a problem possesses some symmetry (e.g. plane, cylindrical or spherical symmetry) then Gauss's law is normally the easiest method to adopt. If, however, there is no symmetry then the calculation must be done using superposition. There are many problems where there is neither symmetry nor knowledge of the actual charge distribution. In such cases, much more elaborate techniques are required and these will be described in later sections.

PROBLEM

P2 Using Gauss's law calculate the electric field E due to:
(*a*) a charge Q distributed uniformly on a spherical surface of radius a;
(*b*) a charge Q distributed uniformly inside a spherical volume of radius a;
(*c*) a large, thin, plane sheet of charge, where the charge density is uniform and equal to q_s in $\mathrm{C\,m^{-2}}$
(*Answer:* (*a*) $Q/4\pi\varepsilon_0 r^2$, (*b*) $Q/4\pi\varepsilon_0 r^2$, (*c*) $q_s/2\varepsilon_0$.)

2.3 ELECTRIC POTENTIAL

The influence set up by electric charges, was described up to now by the electric field **E** or, equivalently, by the electric flux density **D**. Both **E** and **D** are vector quantities and this leads to computational complexity when calculating the electric field due to a number of electric charges.

In this section, another quantity is introduced, which describes the influence set up by charges, but which is not a vector. Its value at any point is given by a signed number—there is no direction associated with it.

Consider a point charge Q as shown in Figure 2.8. A test charge of 1 C is moved along path C from point a to point b as shown. The problem posed is that of finding the amount of work done in moving the test charge from a to b. Clearly, work needs to be done to overcome the Coulomb force between the test charge and Q. It is known from mechanics that moving an object along the path $\Delta\mathbf{l}$ (Figure 2.9) whilst applying a force $\mathbf{F}$ requires an amount of work $\Delta w = F \times \Delta l \times \cos\vartheta$, where ϑ is the angle between the directions of the force and of the path, and Δl is the length of the

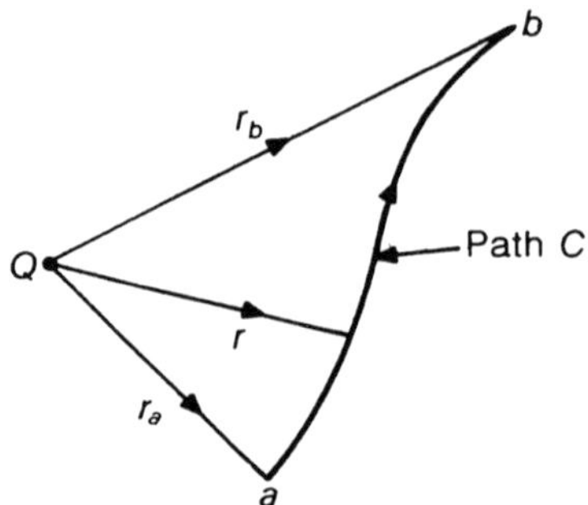

Figure 2.8 Determination of the energy required to move a unit charge from a to b along path C under the influence of charge Q.

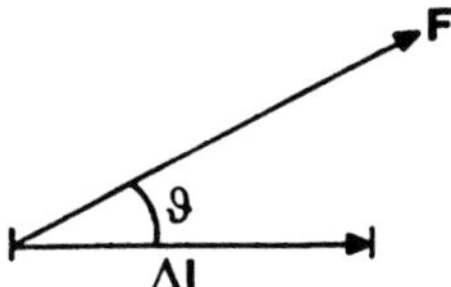

Figure 2.9 Work required to move along **Δl** while applying force **F** is **F·dl** or $F \times \Delta l \times \cos\vartheta$.

path. In vector notation $dw = \mathbf{F}\cdot d\mathbf{l}$ and clearly, since the work depends on ϑ, it may turn out to be either positive or negative. In the former case the experimenter is doing work and in the latter is gaining work. Let us now adapt these ideas to the electrical problem.

The force acting on the test charge is obtained from equation (2.4), $\mathbf{F} = (1\ \text{C}) \times \mathbf{E} = \mathbf{E}$. Hence, the force required to move the charge is $-\mathbf{E}$. The work required to move the charge from a to b may be found by dividing the path C into small segments (Figure 2.10), calculating the work done for each segment, and adding up all contributions:

$$\text{work} = -\Delta l_1 \times E_1 \times \cos\vartheta_1 - \Delta l_2 \times E_2 \times \cos\vartheta_2 - \Delta l_3 \times E_3 \times \cos\vartheta_3 \times \cdots$$

$$= -\sum_n \Delta l_n \times E_n \times \cos\vartheta_n \qquad (2.10)$$

where the summation is over all segments covering the path C from a to b.

As the segment length gets smaller (infinitesimally small), the formula for the work can be written as:

$$\text{work} = -\int_a^b \mathbf{E}\cdot d\mathbf{l} \qquad (2.11)$$

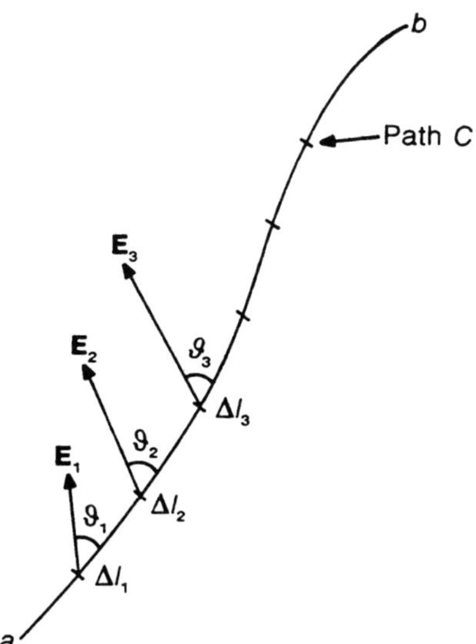

Figure 2.10 Calculation of a 'line integral'.

where the dot product $\mathbf{E}\cdot\mathbf{dl}$ was introduced to indicate the quantity $E \times dl \times \cos\vartheta$. The integral in (2.11) is evaluated over a line joining points a and b, and is known in mathematics as a line integral. It is nothing more than a more compact representation of the computation described by equation (2.10). The details of one segment along the path are as shown in Figure 2.11. The field for a point charge Q is radial and hence $\mathbf{E}\cdot\mathbf{dl} = E \times dl \times \cos\vartheta = E \times dr$ (for small segments). Hence

$$\text{work} = -\int_a^b \mathbf{E}\cdot\mathbf{dl} = -\int_a^b \frac{Q}{4\pi\varepsilon_0 r^2}\,dr = -\frac{Q}{4\pi\varepsilon_0}\int_a^b \frac{dr}{r^2} = -\frac{Q}{4\pi\varepsilon_0}\left(\frac{1}{r_a} - \frac{1}{r_b}\right).$$

The important conclusion drawn from this expression is that, in a static field, the work done in moving a charge is independent of the path followed. Since work depends on initial and final positions only, it should be possible to define a quantity at each point in space r, known as the electric potential $V(r)$, such that

$$\text{work} = -\int_a^b \mathbf{E}\cdot\mathbf{dl} = V(b) - V(a). \tag{2.12}$$

For the particular problem under consideration, the potential at any point a distance r away from Q is

$$V(r) = \frac{Q}{4\pi\varepsilon_0}\frac{1}{r}. \tag{2.13}$$

Hence the work done in moving a test charge of 1 C from a to b is equal to $V(b) - V(a)$. If the unit charge is replaced by a charge q then the work done is $q \times (V(b) - V(a))$. The potential is measured in volts. Moving a positive charge from a higher to a lower potential results in negative work, i.e. a gain in energy for the experimenter. Work is actually done by the experimenter in moving the same charge from low to high potential.

If the potential at a point a distance r_1 from Q_1 and r_2 from Q_2 is required, then, using superposition

$$V(r) = \frac{1}{4\pi\varepsilon_0}\left(\frac{Q_1}{r_1} + \frac{Q_2}{r_2}\right).$$

Hence, calculating the potential V due to a collection of charges is rather easier than

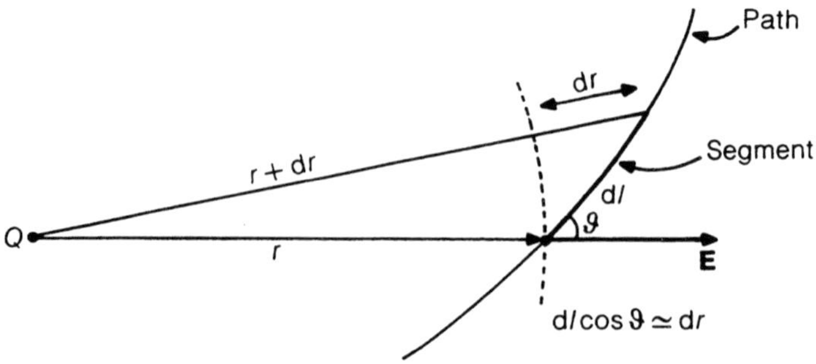

Figure 2.11 Proof that $\mathbf{E}\cdot\mathbf{dl} = E\,dr$.

calculating the electric field **E**, since in the former case it involves addition of numbers whilst in the latter it involves addition of vectors.

Equation (2.12) describes a procedure for calculating potential differences once the electric field is known. It remains to show that the reverse is true, namely, given the potential at any point in space the electric field can be calculated. This calculation is done as follows:

Let us calculate the work done in moving a unit charge from a point with coordinates (x, y, z) to an adjacent point $(x + \Delta x, y, z)$, where Δx is small. Using equation (2.12) and, since movement in the x-direction only is attempted,

$$\text{work} = -E_x \Delta x = V(x + \Delta x, y, z) - V(x, y, z)$$

where E_x indicates the component of the total electric field which is parallel to the x-direction. From this expression:

$$E_x = -\frac{V(x + \Delta x, y, z) - V(x, y, z)}{\Delta x}. \tag{2.14}$$

This equation describes a procedure for calculating the electric field (its component in the x-direction) once the potential is known. Similar expressions may be obtained for the y- and z-components, e.g.

$$E_y = -\frac{V(x, y + \Delta y, z) - V(x, y, z)}{\Delta y}.$$

Readers familiar with calculus will recognise the right-hand side of (2.14) as the derivative of the function $V(x, y, z)$ with respect to x. In (2.14) $V(x, y, z)$ is differentiated with respect to x whilst y and z are treated as constants. This operation on $V(x, y, z)$ is known in mathematics as the partial derivative with respect to x, and it is written as $\partial V(x, y, z)/\partial x$. Hence, in a more compact form:

$$E_x = -\frac{\partial V}{\partial x}, \qquad E_y = -\frac{\partial V}{\partial y}, \qquad E_z = -\frac{\partial V}{\partial z}. \tag{2.15}$$

The magnitude of the field may be obtained from:

$$|E| = \sqrt{E_x^2 + E_y^2 + E_z^2}.$$

Equations (2.15) or (2.14) describe a procedure for calculating the electric field once the potential is known. The magnitude of the E-field is equal to the potential difference per unit length. The direction of the E-field is from points of high to points of low potential. The potential is another way of quantifying the influence of electric charges.

REMARKS

It follows from (2.12) that if a closed path is traced, then:

$$\int_{\text{closed path}} \mathbf{E} \cdot d\mathbf{l} = 0. \tag{2.16}$$

Since the electric field is equal to the potential difference per unit length, $\mathbf{E}\cdot\mathbf{dl}$ represents the potential drop along a short path, and since the integral is equal to the sum of potential drops along a closed path, equation (2.16) is none other than Kirchoff's voltage law. Equation (2.16) is valid for static fields. For time-varying fields, where induction is an important process, (2.16) must be modified.

The expression $V(b) - V(a)$ is the potential difference between points b and a or, equivalently, the potential of point b with respect to a. Hence,

$$V_{ba} = -\int_a^b \mathbf{E}\cdot\mathbf{dl} = \int_b^a \mathbf{E}\cdot\mathbf{dl}.$$

Similarly, the potential of a with respect to b is:

$$V_{ab} = \int_a^b \mathbf{E}\cdot\mathbf{dl} = -V_{ba}.$$

It is important that the reference (point with respect to which potentials are calculated) is clearly stated, otherwise, the sign of the potential (i.e. whether it is higher or lower with respect to the potential of other points) is difficult to ascertain.

Equation (2.13) gives the potential for one important but special case, namely, that of a point charge. It cannot be used when the charge is not a point charge. In such cases, it may be profitable to divide the charge into small elements (which look like point charges) and use superposition to find the total potential. For example, if an electric charge is distributed over a volume of space as shown in Figure 2.12, so that the volume charge density is ρ_v (in $\mathrm{C\,m^{-3}}$), then the potential of a point at a distance r away from the small element of volume dv is $\rho_v\,dv/4\pi\varepsilon_0 r$ and hence the potential at this point due to all the charge is

$$V(r) = \int_{\text{Volume } V} \frac{\rho_v\,dV}{4\pi\varepsilon_0 r}. \tag{2.17}$$

Alternatively, the electric field may be obtained first, by using Gauss's law, and the potential then found from (2.12). Which procedure is adopted depends on the nature of the problem and the experience and ingenuity of the person doing the calculation.

A useful way of visualizing the electric field is to draw lines joining together points

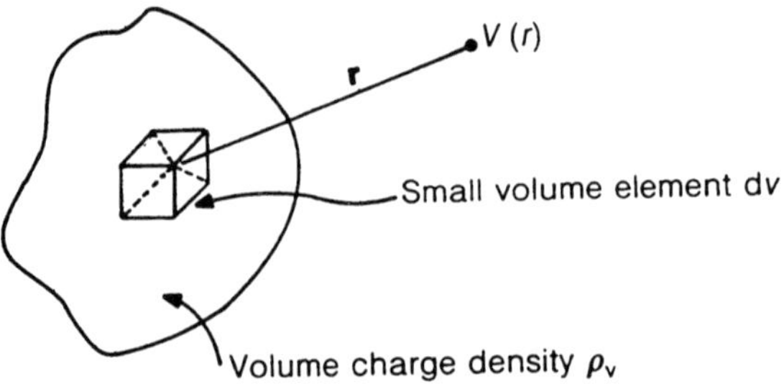

Figure 2.12 Determination of the potential due to a general distribution of charge ρ_v.

of the same potential with respect to some reference point. The lines thus obtained are called *equipotential lines*. Examples of equipotentials will be given in section 2.6.

EXAMPLE E5

Point charges $Q_1 = 10^{-6}$ C and $Q_2 = -5 \times 10^{-6}$ C are placed at $x = 0$ and $x = 0.1$ m respectively, as shown in Figure E5. Calculate:
(*a*) the potential difference $V_{P_1P_2}$;
(*b*) the work done by an experimenter moving a charge $q = -10^{-7}$ C from P_1 to P_2. Indicate clearly whether the experimenter is doing or gaining work.

$Q_1 = 10^{-6}$ C — P_1 — P_2 — $Q_2 = -5 \times 10^{-6}$ C

0 — 0.02 — 0.04 — x (m) — 0.1

Figure E5

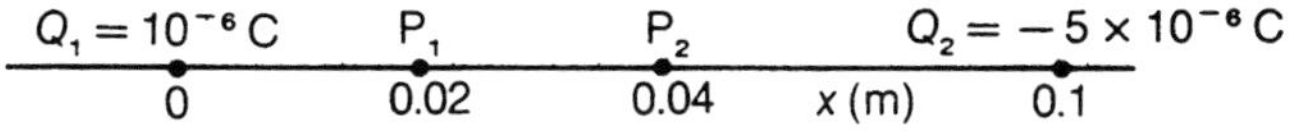

Solution

(*a*) The potential at P_1 and P_2 is due to Q_1 and Q_2 and may be found by superposition. For point charges (2.13) may be used, hence:

$$V_{P_1} = \frac{Q_1}{4\pi\varepsilon_0 \times 0.02} + \frac{Q_2}{4\pi\varepsilon_0 \times 0.08} = -\frac{12.5 \times 10^{-6}}{4\pi\varepsilon_0}$$

$$V_{P_2} = \frac{Q_2}{4\pi\varepsilon_0 \times 0.04} + \frac{Q_2}{4\pi\varepsilon_0 \times 0.06} = -\frac{58.3 \times 10^{-6}}{4\pi\varepsilon_0}$$

$$V_{P_2P_1} = V_{P_2} - V_{P_1} = -412\,\text{kV}.$$

P_2 is at a lower potential compared to P_1.
(*b*) Work done to move q from P_1 to P_2 is:

$$q(V_{P_2} - V_{P_1}) = (-10^{-7})(-412 \times 10^3) = 41.2\,\text{mJ}.$$

The work is found to be positive, hence the experimenter does work. This result makes sense, since a negative charge q is moved to a lower potential.

EXAMPLE E6

A point charge Q is placed at a point (1, 0) as shown in Figure E6. (*a*) Obtain the electric field as a function of Q at the point (0, 1). (*b*) Obtain the function $V(x, y)$ describing the potential due to Q and hence derive an expression for the electric field at (0, 1). Confirm that the result is identical to that found in (*a*).

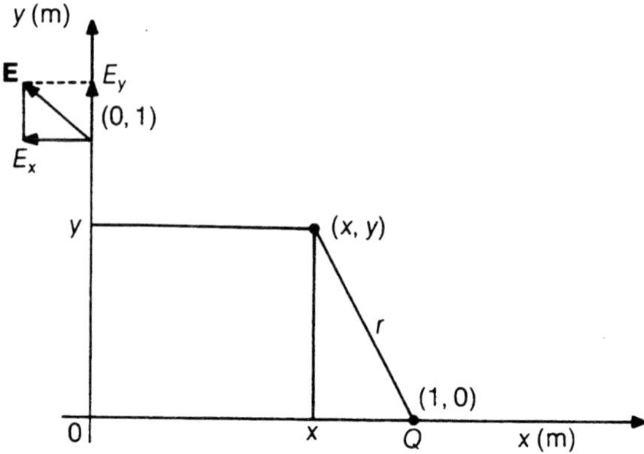

Figure E6 Calculation of the potential $V(x, y)$ due to Q.

Solution

(a) The field at (0, 1) is $E = Q/4\pi\varepsilon_0(1^2 + 1^2) = Q/8\pi\varepsilon_0$, it is parallel to the line joining Q with (0, 1) and directed away from Q.
(b) The potential $V(x, y)$ at (x, y) due to Q is $V(x, y) = Q/4\pi\varepsilon_0 r$ where r is the distance between Q and (x, y) as shown in Figure E6. From this figure $r^2 = y^2 + (1 - x)^2$, hence

$$V(x, y) = \frac{Q}{4\pi\varepsilon_0}\frac{1}{\sqrt{y^2 + (1 - x)^2}}.$$

The component of the electric field in the x-direction is

$$E_x(x, y) = -\frac{\partial V(x, y)}{\partial x} = -\frac{Q}{4\pi\varepsilon_0}(-0.5)[y^2 + (1 - x)^2]^{-3/2}2(1 - x)(-1)$$

$$= \frac{-Q}{4\pi\varepsilon_0}[y^2 + (1 - x)^2]^{-3/2}(1 - x)$$

where the following expression was used for the derivative:

$$[f(x)^{-1/2}]' = -0.5 f(x)^{-3/2} f'(x).$$

Similarly

$$E_y(x, y) = \frac{Q}{4\pi\varepsilon_0}[y^2 + (1 - x)^2]y.$$

Evaluating E_x, E_y at (0, 1) gives:

$$E_x = -\frac{Q}{4\pi\varepsilon_0}2^{-3/2} \qquad E_y = \frac{Q}{4\pi\varepsilon_0}2^{-3/2}.$$

Hence

$$|E| = \sqrt{E_x^2 + E_y^2} = \frac{Q}{8\pi\varepsilon_0}.$$

The E_x, E_y components are shown in Figure E6 in agreement with results in (a).

PROBLEMS

P3 (*a*) Using formulae for the potential due to a point charge, obtain a formula for the potential of point A with respect to point B, due to the point charge $+Q$ as shown in Figure P3.

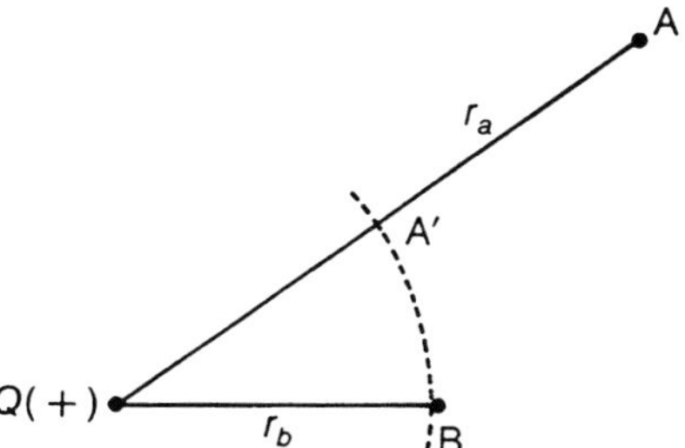

Figure P3

(*b*) Obtain a formula for the electric field, and calculate the potential difference V_{AB} by integration. Confirm that the results are identical to those found in (*a*).
Hint: choose an integration path that makes the calculation easy. Try path AA′B shown in Figure P3.

$$\left(Answer\text{: } V_{AB} = \frac{Q}{4\pi\varepsilon_0}\left(\frac{1}{r_a} - \frac{1}{r_b}\right).\right)$$

P4 Two point charges Q_1 and Q_2 are placed as shown in Figure P4. A charge q is moved from point A to point B. Calculate the work done in moving this charge. Would the experimenter moving this charge need to do work?
(*Answer*: -44.9 μJ, energy gained.)

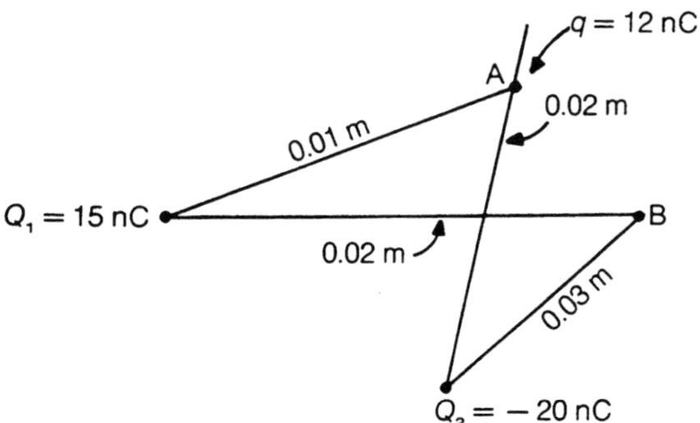

Figure P4

2.4 DIELECTRIC MATERIALS

The electric field, as described up to now, was set up by electric charges surrounded by vacuum or air, i.e. by what is described as free space. However, in engineering applications this is seldom the case. Charges normally reside on material bodies which in turn are surrounded by air or other gases, liquids (such as insulating oils) and solids (such as the PVC insulation on electric wires). It is therefore necessary to extend the ideas and techniques described in the previous sections, to include electric fields in the presence of materials. All materials contain electric charges. However, in some of them these charges are free to move (conductors) whilst in others they are restricted in their movement (insulators or dielectric materials).

In this section dielectric materials are examined, and more specifically we will be concerned with ideal dielectrics, namely those materials which do not contain free charges. Real dielectric materials always contain small amounts of free charge—they are not ideal. Non-ideal (or lossy) materials are studied in the next chapter.

Consider an ideal dielectric containing a large number of atoms as shown in Figure 2.13(*a*). In the absence of an applied electric field, the negative and positive charges are not only equal in magnitude, but are also symmetrically distributed. When an electric field **E** is applied, as in Figure 2.13(*b*), the charge distribution is distorted as shown. This causes a relative separation between positive and negative charges and the material is said to be polarized. In Figure 2.14 a situation is depicted, where half the space is occupied by an ideal dielectric, and the rest by vacuum. The electric field in the vacuum region is **E**, whilst inside the dielectric it is **E**′, possibly different in magnitude to **E**. The problem posed is to relate the electric field **E**′ inside the material to the electric field **E** outside.

At the interface and on the material side, there is some negative charge due to the polarization of the material. Let us call the amount of this charge per m^2 of the interface the *surface charge density*, $-\sigma_b$, where the subscript b stands for bound charge to distinguish it from free charges. Visualize now a cylindrical surface of base area S and height h as shown in Figure 2.14, and apply Gauss's law on the surface. The height can be made small (infinitesimally small) so that the contribution of the curved surface to the Gaussian integral can be made negligible. The contribution of the base of the cylinder and on the vacuum side is $-\varepsilon_0 ES$ (negative, since **E** is pointing into the volume bounded by the surface, and $\cos\vartheta = 1$ since **E** is perpendicular to the surface). Similarly, the contribution from the material side is $\varepsilon_0 E'S$, and the total charge enclosed is $-\sigma_b S$. Hence Gauss's Law gives

$$\varepsilon_0(E' - E)S = -\sigma_b S \qquad \text{or} \qquad E' = E - (\sigma_b/\varepsilon_0). \tag{2.18}$$

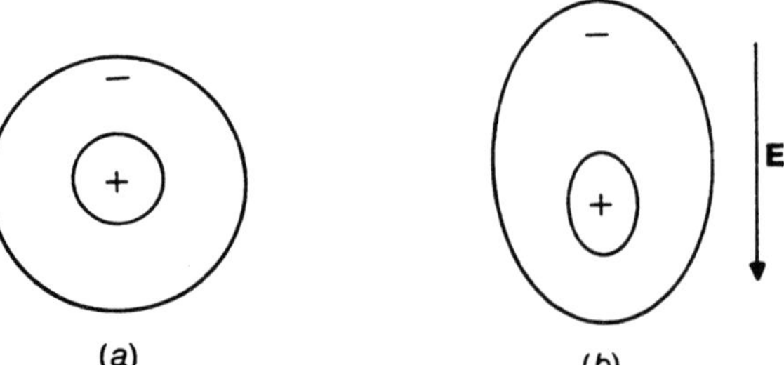

Figure 2.13 The process of 'polarization' in the presence of an electric field **E**.

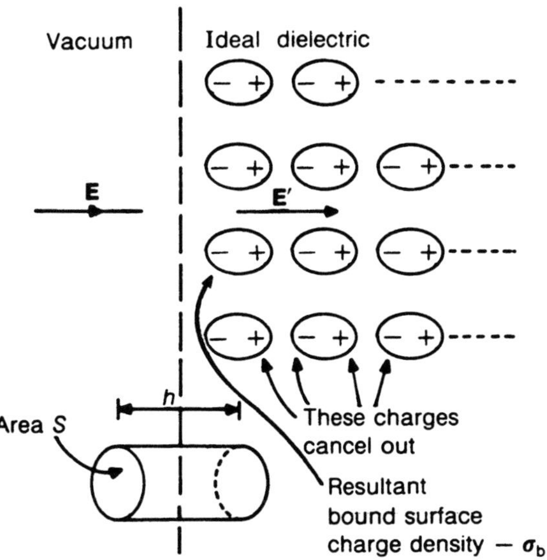

Figure 2.14 Conditions at a vacuum/ideal dielectric boundary.

It is apparent that the polarization reduces the electric field inside the material. A polarization vector $\mathbf{P}$ may be defined having a magnitude σ_b and a direction parallel to $\mathbf{E}$, and then $\mathbf{E}' = \mathbf{E} - \mathbf{P}/\varepsilon_0$.

The electric flux density inside the material is then defined by the expression $\mathbf{D}' = \varepsilon_0\mathbf{E}' + \mathbf{P}$. In most ordinary materials, the polarization is proportional to the electric field inside the material, i.e. $\mathbf{P} = \chi_e\varepsilon_0\mathbf{E}'$, where χ_e is a material constant known as the electric susceptibility. Hence

$$\mathbf{D}' = \varepsilon_0\mathbf{E}' + \chi_e\varepsilon_0\mathbf{E}' = \varepsilon_0(1 + \chi_e)\mathbf{E}' = \varepsilon_r\varepsilon_0\mathbf{E}' = \varepsilon\mathbf{E}'.$$

The quantity ε_r is called the relative permittivity of the material and $\varepsilon(=\varepsilon_r\varepsilon_0)$ its permittivity. To summarize, in free space $\mathbf{D} = \varepsilon_0\mathbf{E}$, whilst inside dielectrics

$$\mathbf{D} = \varepsilon_r\varepsilon_0\mathbf{E} = \varepsilon\mathbf{E}. \tag{2.19}$$

For most liquids $\varepsilon_r \simeq 3$, and in solids $\varepsilon_r \simeq 5$. There are a few exceptions to these typical values namely, water ($\varepsilon_r \simeq 80$) and various titanates ($\varepsilon_r \simeq 10\,000$).

All mathematical models developed so far to describe the electric field in free space can be used in the presence of materials, provided ε_0 is replaced by $\varepsilon = \varepsilon_r\varepsilon_0$. In particular, Gauss's law, as expressed by equation (2.9) is still valid, provided the electric flux density is defined as $\mathbf{D} = \varepsilon_r\varepsilon_0\mathbf{E} = \varepsilon\mathbf{E}$ and the total charge enclosed includes free charges only:

$$\int_S \mathbf{D}\cdot\mathbf{ds} = \text{total free charge enclosed.} \tag{2.20}$$

REMARKS

In applying Gauss's law in the problem shown in Figure 2.14, the expression, $\int_S \varepsilon_0 \mathbf{E}\cdot d\mathbf{s}$ = total charge enclosed, was used where total charge meant free and bound charges. In this particular problem only bound charges were present. In engineering applications, it is more convenient to deal with free charges alone, and the effect of bound charges is accounted for by modifying the relationship (2.19) between **D** and **E**. This leads to the general form (2.20) of Gauss's law applicable to all materials. Equation (2.20) should be used under all circumstances with the appropriate value of ε in equation (2.19). There are circumstances where knowledge of the distribution of bound charges is desirable. In such cases, equation $\mathbf{D} = \varepsilon_0 \mathbf{E} + \mathbf{P}$ may be used to obtain σ_b. Examples of such calculations will be given in section 2.6.

The reader may wonder how is it possible to define an electric field $\mathbf{E}'$ inside the material in the direction shown. In a material containing a vast number of atoms the electric field around each atom will exhibit complex spatial behaviour. However, the electric field is, in this case, the average of these microscopic fields calculated over a small volume. It does not show any of the fine irregularities associated with the fine structure of the material. It is called the macroscopic field and its properties are the subject of this chapter.

2.5 CONDITIONS AT THE BOUNDARIES BETWEEN DIFFERENT MATERIALS

The treatment of electric fields inside ideal dielectrics was presented in section 2.4. Before studying in a systematic way changes to the electric field across the boundaries between different materials, it is necessary to examine the structure of the electric field inside ideal conductors.

The cross-section of a perfect (ideal) conductor is shown in Figure 2.15(*a*). Since the conductor is perfect, no potential drop can exist between any two points in its cross-section. Therefore, the electric field (change in potential per unit length) is zero. Any net free charges placed on a perfect conductor will be subject to strong repulsive forces, and hence will migrate to the outer boundary. Inside the conductor, $\mathbf{E} = 0$, and the free charge density is equal to zero. Similar arguments can be employed for the hollow conductor shown in Figure 2.15(*b*). The electric field inside the cavity will be zero. In real conductors these conclusions have to be modified somewhat, but for most applications it can be assumed to a high degree of accuracy that $\mathbf{E} = 0$ inside a conductor. Consider now the *boundary between a perfect conductor and an ideal dielectric*, as shown in Figure 2.16. As already explained, the electric field inside the conductor is zero. The question arises whether any conditions are imposed on the electric field on the dielectric side of the boundary. It turns out that such conditions exist and can be obtained from the two fundamental laws.

Kirchhoff's voltage law $$\int_{\text{closed path}} \mathbf{E}\cdot d\mathbf{l} = 0 \qquad (2.16)$$

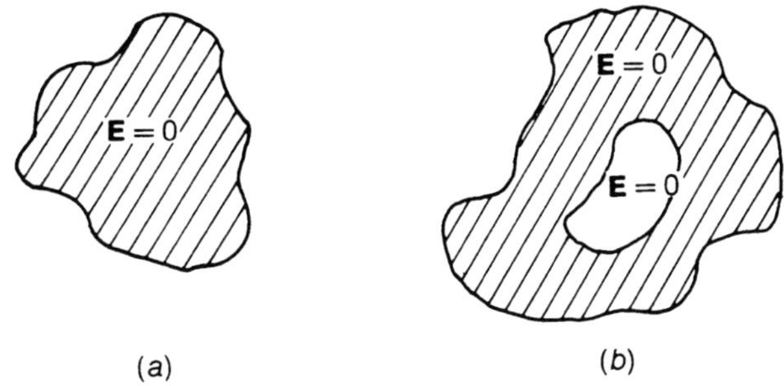

Figure 2.15 The electric field inside perfect conductors.

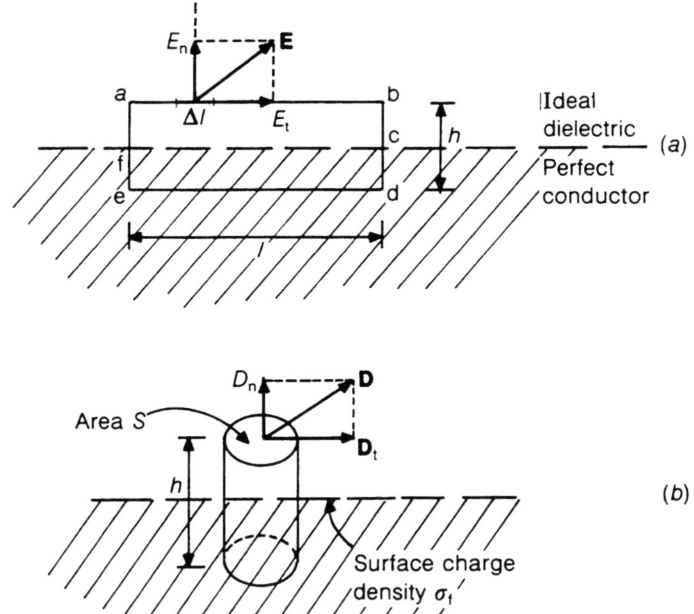

Figure 2.16 Boundary conditions, ideal dielectric/perfect conductor.

$$\text{Gauss's law} \qquad \int_{\substack{\text{closed}\\\text{surface}}} \mathbf{D}\cdot d\mathbf{s} = Q_f. \tag{2.20}$$

To apply (2.16), a closed path abcdefa, as shown in Figure 2.16(*a*), is chosen. The segment ab is chosen short enough to ensure that the electric field does not change significantly along its length. Similarly, since the conditions near the boundary are desired, dimensions h can be made infinitesimally small. No contribution to the integral in (2.16) will be made from path cdef, since $E = 0$ inside the conductor. Similarly, contributions from bc and fa will be negligible for small h. Only path ab will contribute:

$$\int_{\text{abcdefa}} \mathbf{E}\cdot d\mathbf{l} = E_t l = 0,$$

where E_t is the component of the electric field parallel (tangent) to the boundary. It follows from this equation that $E_t = 0$. *The electric field is always normal to the boundary of a perfect conductor.*

In applying (2.20) to this problem, a small cylindrical surface is selected as shown in Figure 2.16(*b*). As before, h can be made arbitrarily small, and S is chosen small enough, to ensure that the flux density **D** does not change significantly over its area. It is also necessary to admit the possibility that free electric charges, of density σ_f (in $\mathrm{C\,m^{-2}}$), may be present on the conductor surface. The portion of the cylinder inside the conductor contributes nothing to the integral in (2.20), since **E** and hence **D** are equal to zero. Similarly, the curved part of the cylinder in the dielectric makes a negligible contribution as h becomes small. The only significant contribution is:

$$\int \mathbf{D}\cdot d\mathbf{s} = D_n S = Q_f = \sigma_f S,$$

where D_n is the component of the electric flux density normal to the interface. Hence, *the electric flux density at the surface of a conductor has the same magnitude as the free charge density on the conductor.* If the charge is positive, the direction of **D** and hence **E** is away from the conductor. For negative charge, **D** and **E** point towards the conductor.

EXAMPLE E7

A perfect conductor has a surface charge density $-2\,\mu\mathrm{C\,m^{-2}}$ and is surrounded by an ideal dielectric of relative permittivity $\varepsilon_r = 3$. Calculate the magnitude and direction of the electric field on the dielectric side of the boundary between the two materials.

Solution

The direction of the field will be normal to the boundary. Since the charge is negative, **D** points towards the conductor and from (2.20) $D = \sigma_f = 2\,\mu\mathrm{C\,m^{-2}}$. Hence, $E = D/\varepsilon_r\varepsilon_0 = 2\times10^{-6}/3\times8.85\times10^{-12} = 75\,\mathrm{kV\,m^{-1}}$.

Let us now consider the *boundary between two ideal dielectrics* as shown in Figure 2.17(*a*). The question posed is the following: given the direction and magnitude of the electric field at a point near the boundary and in, say, dielectric ε_{r1}, is it possible to obtain the electric field near the boundary and in dielectric ε_{r2}? The answer to this question is 'yes' and, as before, the key to obtaining a quantitative answer is in equations (2.16) and (2.20). A closed path abcda is shown in Figure 2.17(*a*) where h, as before, can be made arbitrarily small, and l small

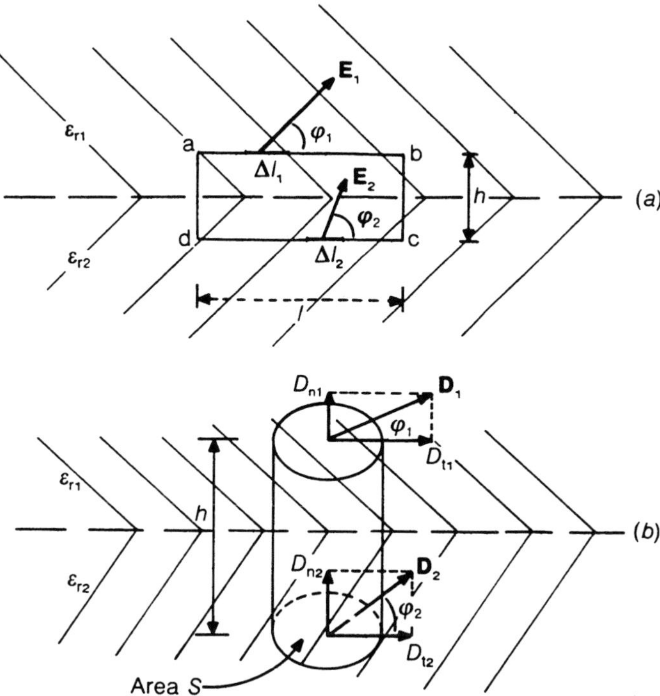

Figure 2.17 Boundary conditions, ideal dielectric/ideal dielectric.

enough so that the electric field along ab remains approximately constant. Applying equation (2.16) gives $lE_1 \cos \varphi_1 - lE_2 \cos \varphi_2 = 0$. The second term is negative, since $\mathbf{E}_2$ points in the opposite sense to that indicated by path abcda. The term $E_1 \cos \varphi_1$ is in fact the component of the electric field in dielectric 1 parallel (or tangent) to the boundary, i.e. E_{t1}. Hence, it follows that, at the boundary between two dielectrics,

$$E_{t1} = E_{t2}. \tag{2.21}$$

Another constraint is obtained by applying (2.20) on the cylindrical surface shown in Figure 2.17(*b*). Only the base and top of the cylinder contribute. Since both dielectrics are ideal, no free charge is enclosed by the cylinder, hence $D_1 \sin \varphi_1 S - D_2 \sin \varphi_2 S = 0$. The second term is negative, since $\mathbf{D}_2$ points into the Gaussian surface. The term $D_1 \sin \varphi_1$ is the component of the electric flux density normal to the boundary, i.e. D_{n1}. Hence, at the boundary between two ideal dielectrics

$$D_{n1} = D_{n2}. \tag{2.22}$$

To summarize, near the boundary between ideal dielectrics, the tangential component of the electric field has the same value on either side of the boundary. Similarly, the normal component of the electric flux density is the same on either side of the boundary.

EXAMPLE E8

In Figure E8, the boundary between two ideal dielectrics is shown. Near the boundary and on the ε_{r1} side, the electric field $\mathbf{E}_1$ is measured and found to be perpendicular to the boundary and equal to $100\,\mathrm{V\,m^{-1}}$.

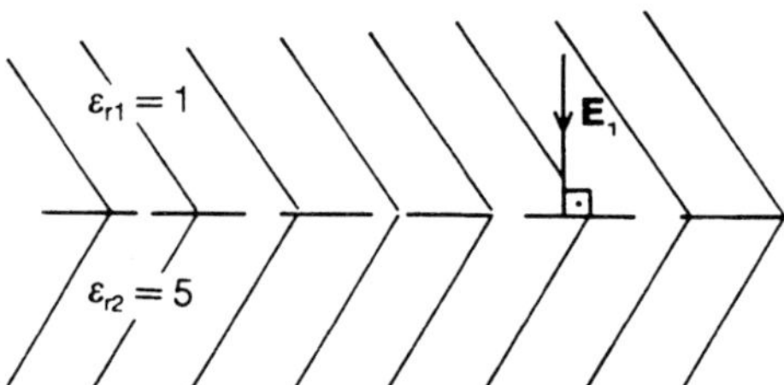

Figure E8

(*a*) What is the direction of the electric field $\mathbf{E}_2$ near the boundary and on the ε_{r2} side?
(*b*) What is the magnitude of $\mathbf{E}_2$?

Solution

(*a*) Apply the boundary condition expressed by (2.21):

$$E_{t1} = E_{t2}.$$

Since $\mathbf{E}_1$ is normal to the boundary, it follows that $E_{t1} = 0$. Hence, E_{t2} is also zero and $\mathbf{E}_2$ must be normal to the boundary.
(*b*) Apply the second boundary condition (2.22), $D_{n1} = D_{n2}$ or, since $\mathbf{D}_1$ and $\mathbf{D}_2$ are normal to the boundary, $D_1 = D_2$. Hence,

$$\varepsilon_{r1}\varepsilon_0 E_1 = \varepsilon_{r2}\varepsilon_0 E_2$$

$$E_2 = \frac{\varepsilon_{r1}}{\varepsilon_{r2}} E_1 = 20\,\mathrm{V\,m^{-1}}.$$

Notice that the electric field is highest in the medium with the smallest dielectric constant. As will be shown later, this conclusion has important implications in the design of electrical insulation.

REMARKS

The boundary conditions derived in this section are very important and should be thoroughly understood. They permit easy calculation of electric fields in practical situations, where the insulation consists of many dielectric layers. Equation (2.22) was obtained from Gauss's law and relies on the fact that, for ideal dielectrics, no free charge is enclosed by the Gaussian surface. In non-ideal dielectrics (lossy dielectrics), some free charge is present and hence (2.22) is not strictly valid.

PROBLEM

P5 Two large blocks of insulating material are separated by a plane boundary as shown in Figure P5. Both materials are ideal with $\varepsilon_{r1} = 2$, and $\varepsilon_{r2} = 5$. The electric field near the boundary and on medium 1 has a magnitude equal to $10\,\mathrm{MV\,m^{-1}}$ and makes an angle $\pi/6$ to the boundary as shown. Calculate the electric field $\mathbf{E}_2$ in medium 2.
(*Answer:* $|E_2| = 8.9 \times 10^6\,\mathrm{V\,m^{-1}}$; 13° to the boundary.)

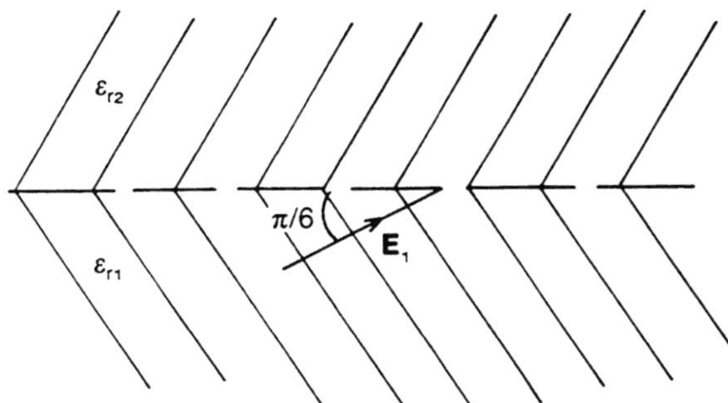

Figure P5

2.6 CALCULATION OF THE ELECTRIC FIELD IN SIMPLE CONFIGURATIONS—CAPACITANCE

It is relatively easy to calculate the electric field due to simple distributions of charge. Whilst doing these calculations to obtain formulae for the electric field, methods for visualizing the structure and strength of the field will also be explored. A number of examples are presented to obtain useful results, and to develop confidence in applying the fundamental laws introduced so far.

Let us start by studying the structure of the electric field around an electric charge Q distributed uniformly on a spherical surface of radius a.

The problem has spherical symmetry (not simply because the charge is on a spherical surface, but also because it is uniformly distributed on it). The electric field may be obtained by applying Gauss's law on a spherical surface of radius r, concentric with the charge, as shown in Figure 2.18(a). On the Gaussian surface the electric field is constant in magnitude and radial in direction (due to spherical symmetry); hence from Gauss's law:

$$D(r)4\pi r^2 = Q \qquad \text{and} \qquad E(r) = D(r)/\varepsilon_0 = Q/4\pi\varepsilon_0 r^2.$$

A test charge q placed near the charge Q will move in a radial direction under a force $\mathbf{F} = q\mathbf{E}$. The electric field and flux lines are radial as shown in Figure 2.18(b)

and are shown symmetrically around the charge Q to indicate the symmetry of the field.

The potential difference between a point A a distance R away from the centre of the spherical charge and the spherical surface (Figure 2.18(c)) is $V_{AB} = \int_{r=R}^{a} \mathbf{E} \cdot \mathbf{dr}$, where the integration is done along the radial direction from A to B. Any other path joining A with the spherical surface could be used, but the path chosen leads to the simplest calculation. Substituting for the electric field gives:

$$V_{AB} = \int_{r=R}^{a} \frac{Q}{4\pi\varepsilon_0 r^2} dr = \frac{Q}{4\pi\varepsilon_0} \int \frac{dr}{r^2} = \frac{Q}{4\pi\varepsilon_0} \frac{r^{-2+1}}{-2+1} \Bigg|_{r=R}^{a} = \frac{Q}{4\pi\varepsilon_0} \left(\frac{1}{R} - \frac{1}{a} \right).$$

Points such as A, a distance R away, are at the same potential. These points lie a on spherical surface which is called the *equipotential surface*. As an example, the equipotential at -500 V with respect to the spherical surface, for the case when $Q = 1$ nC and $a = 1$ cm is obtained by setting $V_{AB} = -500$ V. The radius of this equipotential is found to be $R = 2.25$ cm. The electric flux lines and the equipotentials are perpendicular to each other. This is to be expected, since, if there were a non-zero component of the electric field tangent to the equipotential, this would imply a potential drop along the equipotential.

A common configuration of electrodes is the two parallel conducting plates as shown in Figure 2.19(a). A voltage source V_0 is connected across the plates so that a charge $+\sigma$ (in C m^{-2}) appears on the top plate. From charge conservation,

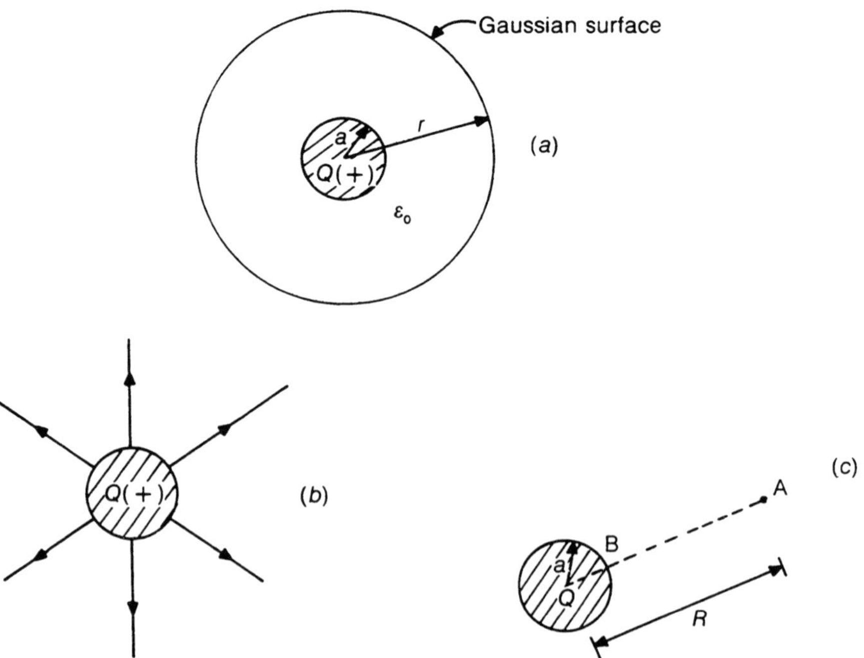

Figure 2.18 Determination of the field of a spherical charge.

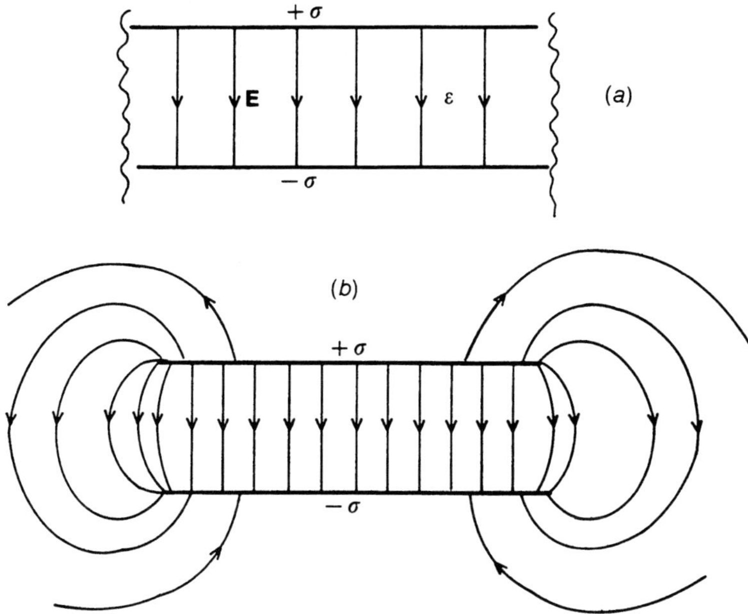

Figure 2.19 The field between the parallel plates of a capacitor.

a charge $-\sigma$ will appear on the bottom plate. The objective is to obtain a formula for the electric field between the two plates.

It can be assumed that the plates are large and that the field away from the edge of the plates is required.

The boundary conditions (section 2.5) require that the electric field **E** must be perpendicular to the conducting plates. Also, the flux density near the plates is equal to the surface charge density, i.e. $D = \sigma$ and hence $E = D/\varepsilon = \sigma/\varepsilon$. The flux lines leave from the positive charge and terminate on the negative charge. The potential difference V between the plates may be obtained from $V = \int \mathbf{E} \cdot \mathbf{dx}$ where, for simplicity, the integration is done along the shortest path from the top to the bottom plate:

$$V = \int E\mathrm{d}x = E \int \mathrm{d}x = Ed = \sigma d/\varepsilon.$$

However, since a source V_0 is connected across the plates, it must be the case that $V_0 = \sigma d/\varepsilon$. Assume now that the surface area of each plate is A. Then $\sigma A = (\varepsilon A/d)V$. The quantity σA is equal to the charge q on each plate. It follows that the ratio of the charge on each plate, over the potential difference between the plates is constant and equal to $\varepsilon A/d$, i.e. it depends only on the geometrical dimensions and the dielectric properties of the insulating material. This quantity (charge on each plate over potential difference) is by definition the *capacitance* C between the two plates:

$$C = \frac{\text{charge on each plate}}{\text{potential difference}}. \tag{2.23}$$

This definition holds true for all conductor configurations, but the formula $C = \varepsilon A/d$ holds true only for two parallel plate conductors of area A, separated by a distance d, and insulated by a dielectric material ε. Figure 2.19(*b*) shows the electric field between the plates and also near the edges. The calculation, as just described, neglected the distortion of the field near the edges. This is normally a reasonable assumption.

EXAMPLE E9

(*a*) A parallel plate capacitor (electrode area A, separation d) is charged as shown in Figure E9(*a*). Plot D, E, and the free/bound charge densities.

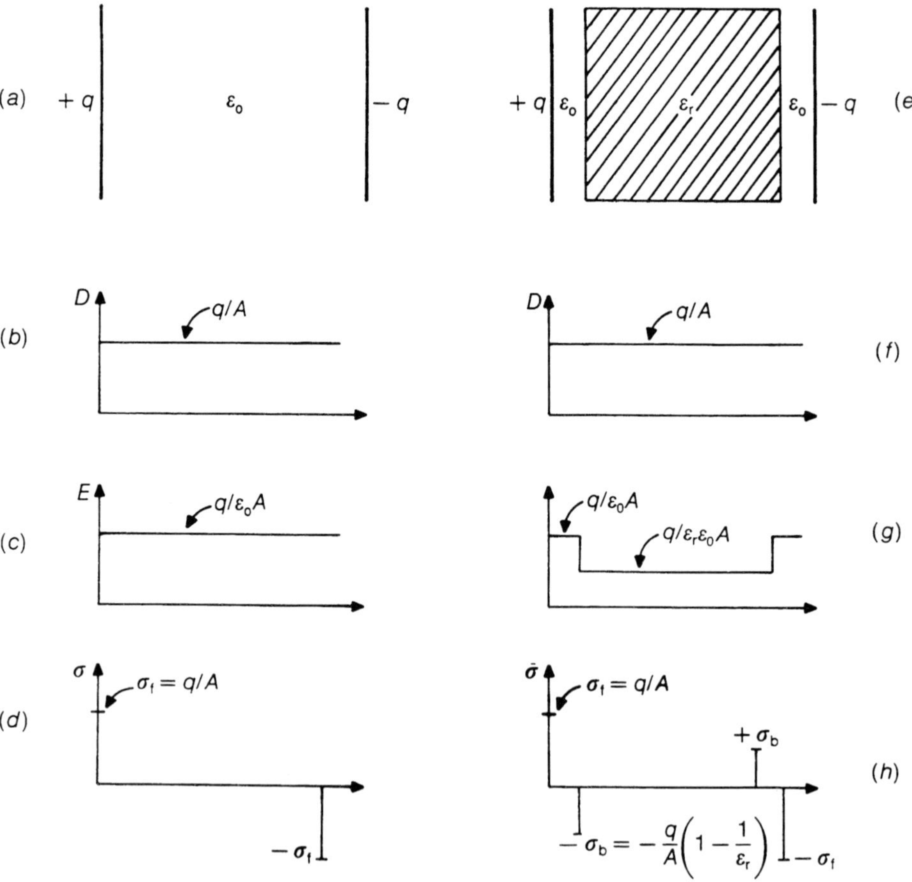

Figure E9 Field quantities in a vacuum (*a*)–(*d*) and ideal dielectric (*e*)–(*h*) insulated capacitor.

(*b*) In the capacitor described in (*a*), a dielectric ε_r is inserted as shown in Figure E9(*e*). Plot D, E, σ_f, σ_b, as in (*a*).

Solution

(*a*) Boundary conditions on the conducting electrodes require that $D = \sigma_f = q/A$ and hence, $E = D/\varepsilon_0 = q/\varepsilon_0 A$. Figures E9(*b, c, d*) are thus obtained. There are no bound charges since the dielectric is ideal.
(*b*) Since there are no connections to the two electrodes, the charge on each electrode must remain unchanged. Boundary conditions near the electrode require that $D = q/A$. At the boundary between the two ideal dielectrics, the normal component of the flux density must be the same in the two media. Hence, D is constant throughout (plot E9(*f*)) and of the same value as in (*a*). The electric field in the vacuum portion is D/ε_0 and in the dielectric $D/\varepsilon_r\varepsilon_0$ as shown in plot E9(*g*). The free charge density in the conducting plates is the same as before. There will now be bound charges on the surface of the dielectric insert. From equation (2.18), $D/\varepsilon_r\varepsilon_0 = D/\varepsilon_0 - \sigma_b/\varepsilon_0$ and hence, $\sigma_b = D(1 - 1/\varepsilon_r)$. Bound and free charge densities are plotted in E9(*h*).

EXAMPLE E10

Calculate the capacitance per metre length between two coaxial conductors as shown in Figure E10.

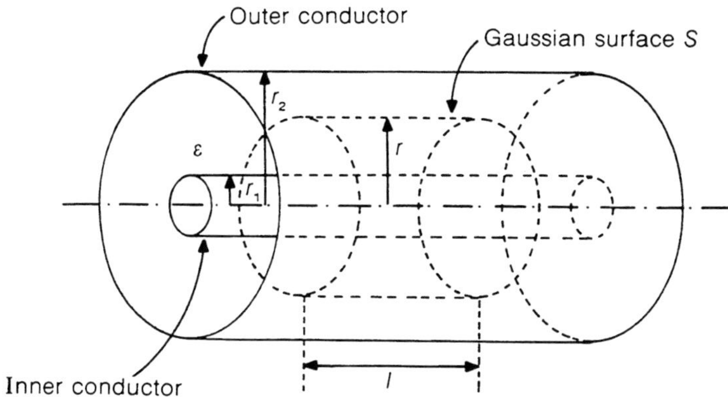

Figure E10 Gaussian surface for determining the field of a cylindrical charge.

Solution

Assume that the coaxial conductors are very long so that edge effects may be neglected. In practice this means that the length of the coaxial conductors far exceeds the radius r_2. Let us assume that by some charging process a charge q_l (in $\mathrm{C\,m^{-1}}$) has been established along the surface of the inner electrode. From symmetry, it is expected that this charge is distributed uniformly around the surface of the inner electrode. Consider next a Gaussian cylindrical surface S, coaxial with the conductors, as shown in Figure E10. From symmetry, $E(r)$ is everywhere radial and of constant magnitude on the curved part of the Gaussian surface. The base and top of S do not contribute to the Gaussian integral, as the normal electric field component there is zero. Hence, $D(r) \times 2\pi r l = q_l l$. Note that only the charge enclosed by S needs to be taken into account. Any charges that may exist on the outer electrode should not be included. Hence, $D(r) = q_l/2\pi r$ or $E(r) = q_l/2\pi\varepsilon r$. These formulae give the electric flux density and the electric field due to a cylindrical distribution of charge, where q_l is the charge per unit length on the cylindrical conductor. The potential difference between the inner and outer electrodes is

$$V = \int_{r=r_1}^{r_2} E(r)\,\mathrm{d}r = \frac{q_l}{2\pi\varepsilon}\ln r\Big|_{r=r_1}^{r_2} = \frac{q_l}{2\pi\varepsilon}\ln\frac{r_2}{r_1}.$$

The capacitance per unit length is defined as the charge per unit length over the potential difference. Hence, $C = q_l/V = (2\pi\varepsilon/\ln(r_2/r_1))$, in F m^{-1}, is the capacitance between the coaxial conductors. To summarize, the *general procedure for calculating capacitance* is:

- Assume a charge q distributed on each electrode. Ascertain whether, from symmetry or other considerations, it is possible to describe how the charge is distributed.
- Apply Gauss's law to calculate the electric field in the space between the electrodes. In fact, it is only necessary to know the electric field along a path joining the two electrodes. The choice of this path must be done carefully to simplify calculations.
- Calculate the potential difference between the two electrodes $V = \int \mathbf{E}\cdot\mathrm{d}\mathbf{l}$, where the integration is done along the path already chosen.
- Calculate the capacitance from $C = q/V$.

EXAMPLE E11

Obtain an expression for the capacitance between two parallel wires (two-wire transmission line) as shown in Figure E11.

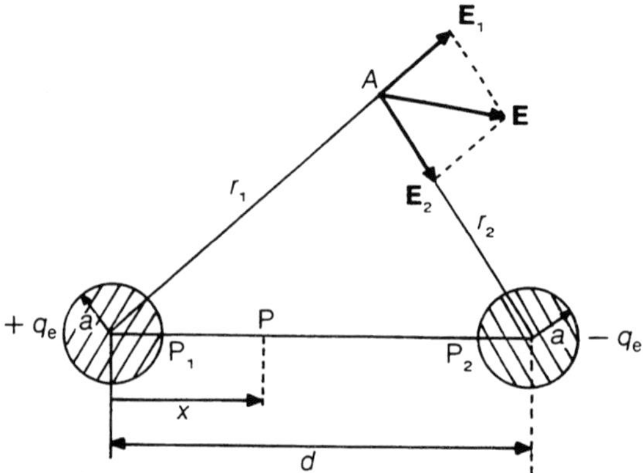

Figure E11 The field due to a two-wire line.

Solution

The procedure outlined in example E10 will be followed.

• Assume that charges $+q_l$ and $-q_l$ (in $\mathrm{C\,m^{-1}}$) are distributed on the two conductors as shown. It is reasonable to assume that charge is distributed uniformly on the surface of each conductor, provided the spacing d is much larger than the radius of the conductors (say $d > 3a$). In such cases the charge distribution on a conductor is not significantly affected by the presence of another.

• The electric field around the two conductors must now be found. At a point A (Figure E11), the electric field $\mathbf{E}$ can be found by superposition, i.e. $\mathbf{E} = \mathbf{E}_1 + \mathbf{E}_2$, where $\mathbf{E}_1$, $\mathbf{E}_2$ are the components of the field due to charges $+q_l$ and $-q_l$ acting alone. For a cylindrical distribution of charge $E_1 = q_l/2\pi\varepsilon r_1$ and $E_2 = q_l/2\pi\varepsilon r_2$ with directions as shown. This result implies a uniform distribution on each conductor. If this assumption is not justified, $\mathbf{E}_1$ and $\mathbf{E}_2$ must be worked out differently using more advanced techniques. The total electric field $\mathbf{E}$ at point A can thus be found. However, only the field along a path joining the two conductors is required. The simplest path is $\mathrm{P_1PP_2}$ because, on that path, both field contributions act in the same direction. Thus at P:

$$E(x) = \frac{q_l}{2\pi\varepsilon x} + \frac{q_l}{2\pi\varepsilon(d-x)}.$$

The first term is due to $+q_l$ and is directed from $\mathrm{P_1}$ to P. The second term is due to $-q_l$ and its direction is from P to $\mathrm{P_2}$. Hence, they add together.

• The potential difference between the two conductors can be found by calculating $\int E_{(x)}\,\mathrm{d}x$ along a path starting from the surface of one conductor (point $\mathrm{P_1}$) and finishing on the other conductor (point $\mathrm{P_2}$):

$$V = \int_{x=a}^{d-a} E(x)\,\mathrm{d}x = \int \frac{q_l}{2\pi\varepsilon}\left(\frac{1}{x} + \frac{1}{d-x}\right)\mathrm{d}x = \frac{q_l}{2\pi\varepsilon}[\ln x - \ln(d-x)]\Big|_{x=a}^{d-a}.$$

• By definition, the capacitance per unit length between the two conductors is

$$C = \frac{q_l}{V} = \frac{\pi\varepsilon}{\ln[(d-a)/a]}\ (\mathrm{F\,m^{-1}}).$$

EXAMPLE E12

For the configuration shown in Figure E11, sketch out the equipotentials.

Solution

Let us find the potential of point A with respect to midpoint O shown in Figure E12. From superposition

$$\begin{aligned} V_{\mathrm{AO}} &= V_{\mathrm{AO}}(\text{due to } +q_l) + V_{\mathrm{AO}}(\text{due to } -q_l) = (V_{\mathrm{AB}} + V_{\mathrm{B'O}})(\text{due to } +q_l) \\ &\quad + (V_{\mathrm{AC}} + V_{\mathrm{C'O}})(\text{due to } -q_l). \end{aligned}$$

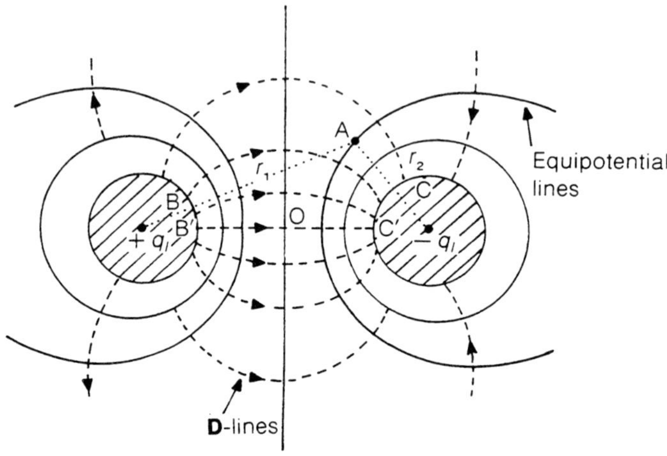

Figure E12 The electric field and equipotential lines around a two-wire line.

From the symmetry of the problem $V_{B'O} = -V_{C'O}$, hence

$$V_{AO} = V_{AB}(\text{due to } +q_l) + V_{AC}(\text{due to } -q_l).$$

But

$$V_{AB}(\text{due to } +q_l \text{ only}) = \int_{r=r_1}^{a} \frac{q_l}{2\pi\varepsilon r}\,dr = \frac{q_l}{2\pi\varepsilon}\ln\frac{a}{r_1}$$

and similarly

$$V_{AC}(\text{due to } -q_l \text{ only}) = -\frac{q_l}{2\pi\varepsilon}\ln\frac{a}{r_2}.$$

Hence

$$V_{AO} = \frac{q_l}{2\pi\varepsilon}\ln\frac{r_2}{r_1}.$$

It follows that points of constant potential must have the same ratio of r_2/r_1. They lie on circles as shown in Figure E12. In three dimensions, the equipotentials are cylinders perpendicular to the paper and intersecting along the circles shown. The special case $r_2/r_1 = 1$ corresponds to an equipotential plane passing from point 0.

EXAMPLE E13

A long cylindrical wire of radius a, runs parallel and at a height h to a large perfectly conducting plane. Calculate the capacitance per unit length between the wire and the plane.

Solution

The configuration of conductors for this problem is shown in Figure E13(a). Following normal procedure, it is postulated that charges $+q_l$ and $-q_l$ (in $\mathrm{C\,m^{-1}}$)

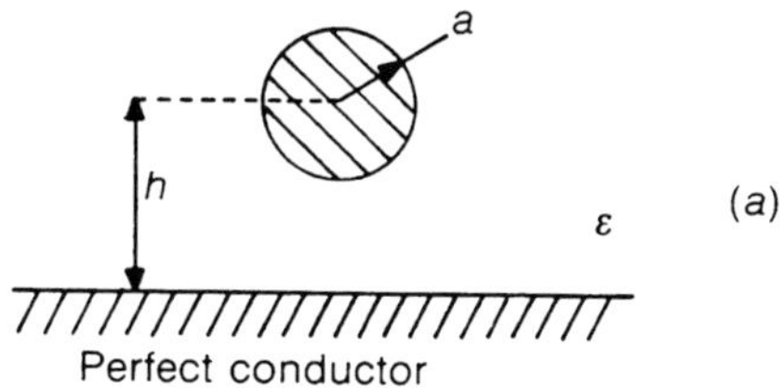

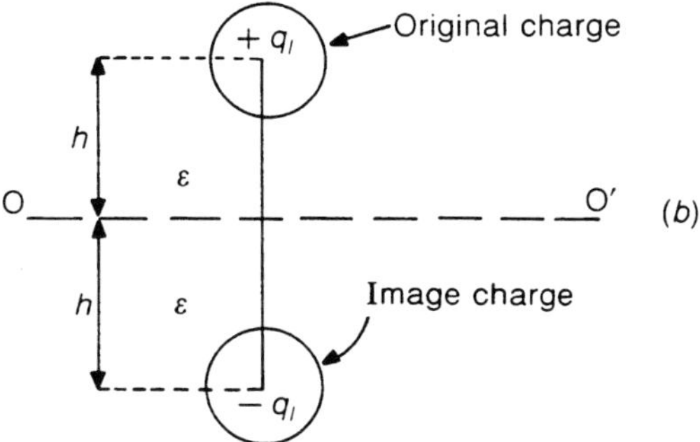

Figure E13 (*a*) Conductor above earth problem and (*b*) its image equivalent.

are established on the wire and plane respectively. The next step is the calculation of the electric field and potential difference between wire and plane. Finally, the capacitance per unit length is calculated by dividing q_l by the potential difference. The difficulty in following this procedure lies in the calculation of the electric field. Although it is a good approximation to assume that $+q_l$ is uniformly distributed on the wire (provided $h \gg a$), it is a very poor approximation to specify that $-q_l$ is uniformly distributed on the plane. Intuitively one expects that on the plane and directly under the wire, substantial negative charge will be established, whilst sufficiently far from this point, the negative charge will have a very small value. Without knowing how the charge is distributed on the plane, it is not clear how the electric field can be calculated. Without an obvious symmetry to the problem, it is not profitable to use Gauss's law. There are many similar problems where, whilst Gauss's law still applies, it is not easy to implement it. In such cases, resort must be made to more powerful computational techniques which are described briefly in Part II. Alternatively, other techniques must be sought which, whilst remaining computationally simple, extend computational capability to problems without an obvious symmetry. Such a technique is the *method of images*. It is based on the following premise: if a solution to a problem is found (by intuitive or other means) which satisfies the basic laws and also conditions at the boundaries of the problem, then this is the true and only solution to the problem. A rigorous proof of this statement can be given, but suffice it to say that, if it were not true, it would mean that more than one solution to a fully defined problem would be possible—a state of affairs at variance with practical experience.

Let us now see how to tackle the problem of Figure E13(a). Consider the problem shown in Figure E13(b) which is identical to that studied in examples E11 and E12. The plane perpendicular to the paper and passing from the line OO′ is an equipotential plane. Hence, the field above this plane OO′, calculated for the configuration in (b), satisfies all the basic laws and in addition it conforms to the boundary conditions of the configuration in (a). Hence, the field above OO′ in (b), is the solution to the problem in (a). The effect of the distributed charge on the plane in (a), is described by the fictitious or image charge in (b). The capacitance between the two conductors in (b) was calculated in example E11 and found to be equal to $\pi\varepsilon/\ln[(2h-a)/a]$ in F m^{-1}. The capacitance between the wire and the plane is equal to $2\pi\varepsilon/\ln[(2h-a)/a]$ in F m^{-1}, as a direct calculation will confirm. Since usually $h \gg a$, to a good approximation the capacitance in $2\pi\varepsilon/\ln(2h/a)$.

The method of images can be applied to many practical problems, such as a conductor near a conducting corner or inside a conducting box.

In cases where the assumption that $+q_l$ is distributed uniformly on the wire is not valid, (i.e. when h/a is not large), more sophisticated techniques using images can be used to give accurate results.

PROBLEMS

P6 Using Gauss's law, calculate the electric field due to a charge σ (in C m^{-2}) deposited on the surface of a semi-infinite block of perfectly conducting material (Figure P6).
(*Answer*: $E=\sigma/\varepsilon$.)

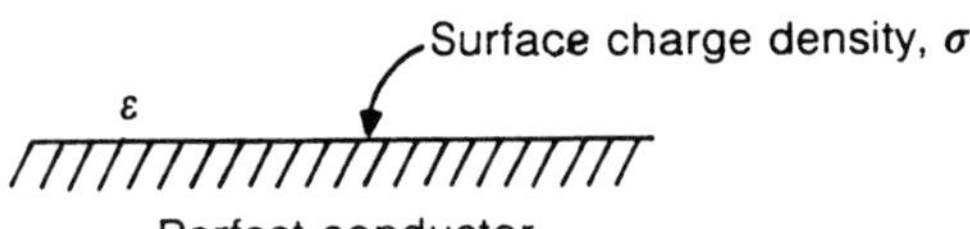

Figure P6

P7 A cylindrical charge is uniformly distributed on a long cylindrical surface of radius $a=5$ cm, so that there are 6 μC per metre length of the cylinder. Sketch the electric flux lines on a plane perpendicular on the cylinder, representing 1 m length of the cylinder, so that there is one flux line per microcoulomb of charge on the cylindrical surface.

In the same diagram sketch the equipotential lines corresponding to potentials of -20 kV and -40 kV with respect to the surface of the cylinder.
(*Answer:* Six symmetrically placed lines, radii of equipotentials 6.02 cm, 7.24 cm.)

P8 A plane capacitor is formed between plates A and B as shown in Figure P8 and is charged by a source to a potential $V_0=10$ V. Between the two plates there are two ideal dielectrics as shown. Edge effects may be neglected. Plot diagrams showing how the electric field E, the electric flux density D, and the surface charge density vary in the space between the plates. Make sure to distinguish between

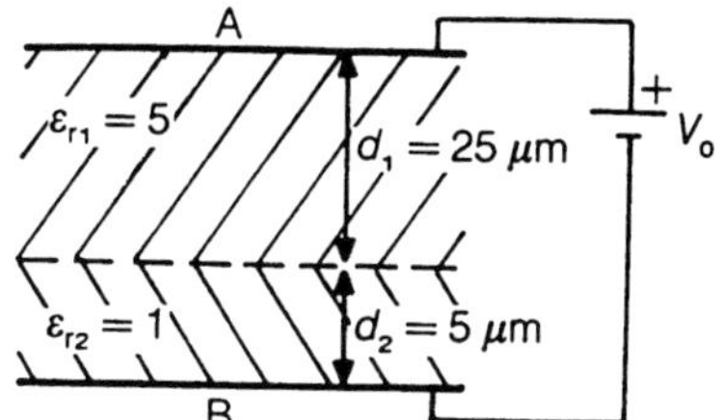

Figure P8

free and bound charge densities. Calculate the capacitance per m^2 for this system. (*Answer:* $0.885\ \mu F\ m^2$).

P9 Two charge layers of thickness a are arranged as shown in Figure P9. The volume charge density is constant throughout each layer and equal to $-\rho_v$ for $-a < x < 0$ and $+\rho_v$ for $0 < x < a$. Assume that outside the double layer the electric field is zero.

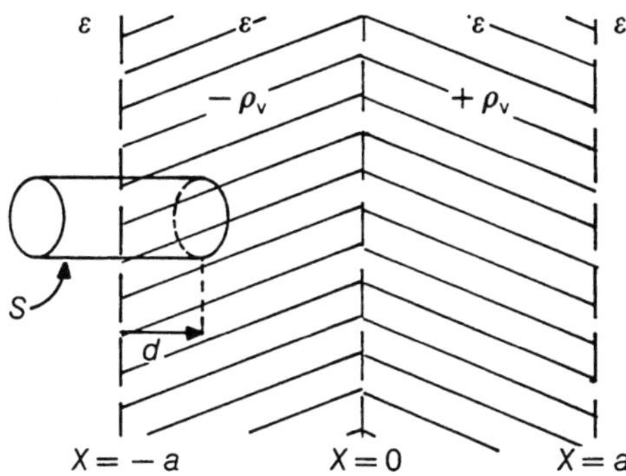

Figure P9

(*a*) Using Gauss's law, derive formulae for the electric field $E(x)$ inside the double layer ($-a < x < a$) and plot $E(x)$ versus x.
(*b*) Derive a formula for the potential difference $V(x)$ using the potential at $x = 0$ as the reference. Sketch $V(x)$ versus x.
Hint: use the Gaussian surface S shown, for values of d ranging from 0 to $2a$.
(*Answer:* peak electric field is $-\rho_v a/\varepsilon$ at $x = 0$ and decreases linearly, $V(x) = -\rho_v(x^2/2 - ax)/\varepsilon$.)

P10 Derive an expression for the capacitance between two conducting concentric spheres of radii r_1, r_2 ($r_2 > r_1$). Derive an expression for the capacitance of a single isolated sphere of radius r_1.
Hint: use the result for the two spheres and take limit $r_2 \to \infty$.
(*Answer:* $C = 4\pi\varepsilon/(1/r_1 - 1/r_2)$, $C = 4\pi\varepsilon r_1$).

P11 Two long conducting cylinders are separated by two layers of perfect dielectric as shown in Figure P11.

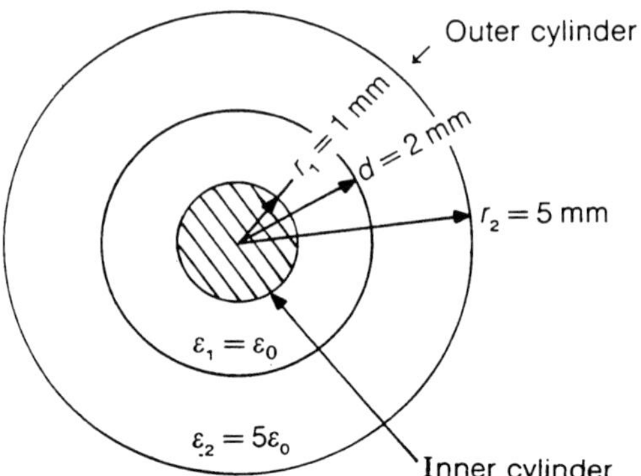

Figure P11

(a) Derive formulae for the electric field between the two cylinders.
(b) Calculate the total capacitance per metre length of the cylinders.
(c) Calculate the maximum value of the field E between the cylinders when a potential difference equal to 100 V is established between the two coaxial electrodes. (*Answer:* (a) first layer $E = q/2\pi\varepsilon_1 r$, second layer $E = q/2\pi\varepsilon_2 r$, ($b$) $C = 63.4\,\text{pF m}^{-1}$, $E_{\text{max}} = 114\,\text{kV m}^{-1}$.)

P12 A long conducting cylinder of radius a is running parallel and at a height h ($h \gg a$) above a large conducting plane as shown in Figure P12.

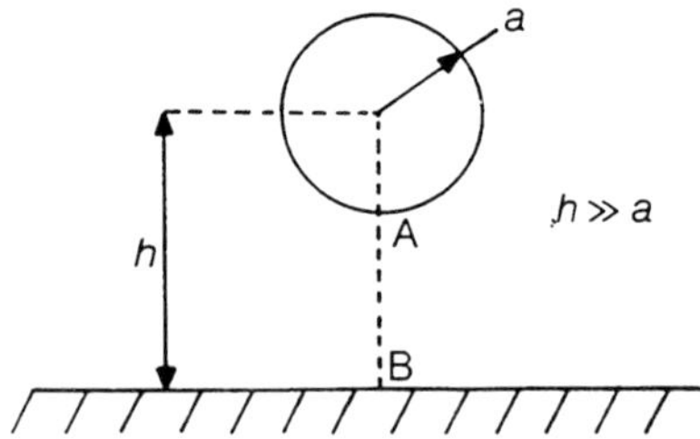

Figure P12

(a) Calculate the capacitance per unit length between the cylinder and plane.
(b) Calculate the electric field E and surface charge density at points A and B, when a potential difference V_0 is established between the cylinder and the plane. (*Answer:* (a) $C = 2\pi\varepsilon/\ln[(2h - a)/a]$
(b) $E_A = V_0(1/a + 1/2h)/\ln[(2h - a)/a]$, $\sigma_A = \varepsilon E_A$ $E_B = 2V_0/h\ln[(2h - a)/a]$, $\sigma_B = \varepsilon E_B$.)

P13 A long cylindrical conductor runs parallel to a perfect earth as shown in Figure P12. Due to a large thunder cloud above the earth, a uniform electric field E_0 pointing upwards is established. The conductor is earthed by a thin wire which produces a negligible disturbance to the field. Calculate the charge per unit length induced on the conductor.
Hint: Assume a charge q_l on the conductor and calculate the total value of the

electric field along the line AB due to q_l, its image, and E_0. Then, since the conductor is earthed, demand that $V_{AB} = 0$.
(*Answer:* $q_l = E_0 h[2\pi\varepsilon_0/\ln(2h/a)]$.)

P14 A charge is distributed uniformly along a long cylindrical conductor running parallel to a conducting corner, so that there are q_l C m^{-1} on the conductor (Figure P14). Obtain an expression in terms of q_l for the force (in N m^{-1}) acting on the charge. Find the direction of this force.
(*Answer:* $|F| = 7.25 \times 10^{11} q_l^2$ N m^{-1}.)

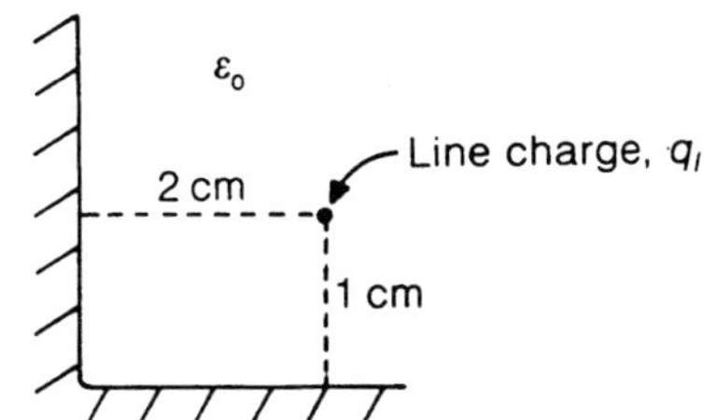

Figure P14

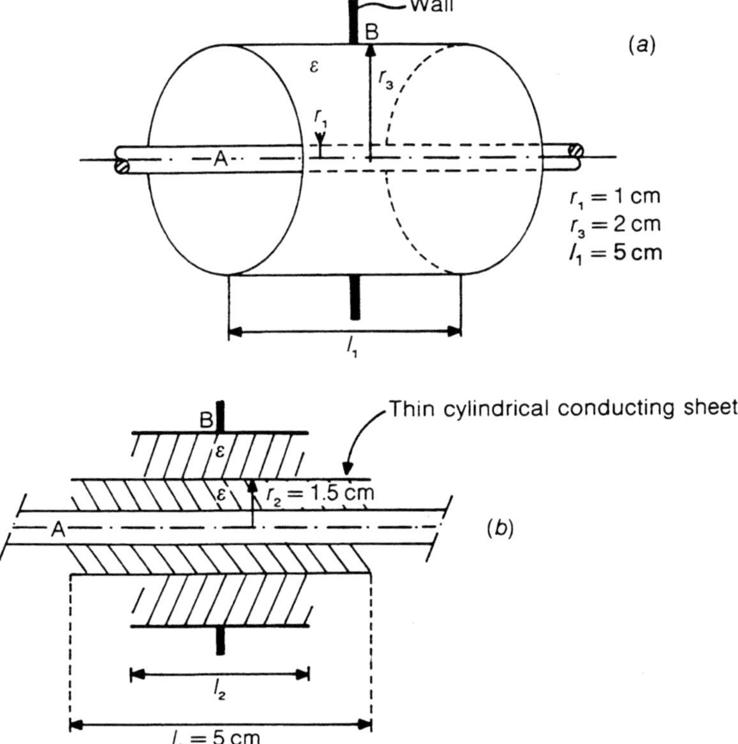

Figure P15 (*a*) Cylindrical conductor A through a cylindrical wall B. (*b*) same configuration but with an intermediate floating cylindrical conducting sheet (capacitor graded bushing).

P15 (*a*) A cylindrical conductor A is insulated from a conducting wall B as shown in Figure P15(*a*). A potential difference V_0 is applied between A and B. Neglecting all fringe effects calculate the maximum electric field in the dielectric. (*b*) An alternative method of insulation is shown in Figure P15(*b*). Show that if a thin conducting layer is added halfway between the insulation, the length l_2 may be chosen in such a way that the maximum electric field is the same in both insulating layers and smaller compared to the value obtained in (*a*).
Hint: the two insulating layers form in effect two coaxial capacitors in series. This has implications on the value of electric charge on each capacitor plate!
(*Answer:* (*a*) $E_{max} = 1.442\ V_0$, (*b*) $E_{max} = 1.193\ V_0$ when $l_2 = 3.33$ cm)

P16 A long cylindrical conductor is placed above a conducting plane as shown in Figure P16(*a*).

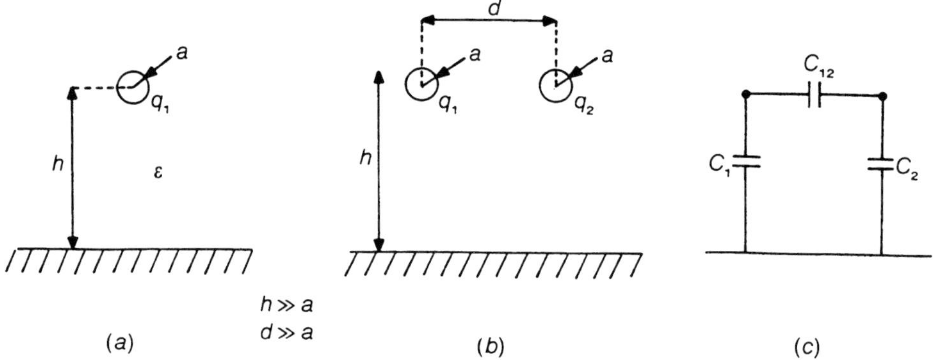

Figure P16

(*a*) Find the potential of the conductor with respect to the plane (assume a charge q_1 per metre length on the cylindrical conductor).
(*b*) A second conductor (charge q_2 per metre) is introduced as shown in Figure P16(*b*). Using the method of images, find the potential of conductor 1 (with respect to the plane) due to conductor 2.
Hint: use superposition.
(*c*) Find the potential of conductor 1 (with respect to the plane) due to both q_1 and q_2.
Hint: use superposition.
(*d*) For the case when $d = 8a$, $h = 3a$ and the conductors are in a medium with $\varepsilon_r = 2$, calculate the p coefficients in the following expression.

$$V_1 = p_{11}q_1 + p_{12}q_2 \qquad V_2 = p_{21}q_1 + p_{22}q_2.$$

Solve these equations to find the c coefficients in the expressions below:

$$q_1 = c_{11}V_1 + c_{12}V_2 \qquad q_2 = c_{21}V_1 + c_{22}V_2.$$

(*e*) Rearrange the last set of equations to show that the circuit in Figure P16(*c*) is the

equivalent circuit for the system of conductors, and calculate the self (C_1, C_2) and mutual (C_{12}) capacitances.
(*Answer:* (*a*) $(q_1/2\pi\varepsilon)\ln(2h/d)$ (*b*) $(q_2/2\pi\varepsilon)\ln(D/d)$ where D is the distance between conductor 1 and the image of conductor 2
(*c*) $V_1=(q_1/2\pi\varepsilon)\ln(2h/a)+(q_2/2\pi\varepsilon)\ln(D/d)$
(*d*) $c_{11}=c_{22}=63\,\text{pF m}^{-1}$, $c_{12}=c_{21}=-7.8\,\text{pF m}^{-1}$
(*e*) $C_1=C_2=c_{11}+c_{12}=55\,\text{pF m}^{-1}$, $C_{12}=-c_{12}=7.8\,\text{pF m}^{-1}$.

2.7 ENERGY STORAGE IN A SYSTEM OF CONDUCTORS

The usefulness and application of electric field systems hinges on their ability to store energy. It is therefore necessary to study energy storage in a system of charged conductors.

The basic ideas will be developed with reference to the system of two parallel conducting plates shown in Figure 2.20. With the two plates initially uncharged, a battery V is connected as shown. The action of the battery is to transfer charge from one plate to the other so that, eventually, there is charge $+Q$ on the top plate. The charge on the bottom plate will then be $-Q$ (charge conservation). The problem posed is to find the amount of energy, if any, stored in the two-electrode system. Let us assume that at some stage during charging, a charge q has been transfered, and that at that moment the potential difference between the plates is v. Obviously, $q=Cv$ where C is the capacitance of the two plates. The amount of work that must be done by the source to transfer an additional amount of charge dq is then d$w=v\,$dq (since potential is work per unit charge), excluding any energy dissipated as heat in the resistance R. The energy dw must be stored in the electrode system (energy conservation). To find the total energy stored when the total charge Q has been transferred, all that is required is to add up all contributions dw

$$W=\int_{q=0}^{Q} v\,\mathrm{d}q=\int_{q=0}^{Q}\frac{q}{C}\,\mathrm{d}q=\left.\frac{q^2}{2C}\right|_{q=0}^{Q}=\frac{Q^2}{2C}=\tfrac{1}{2}CV^2=\tfrac{1}{2}QV$$

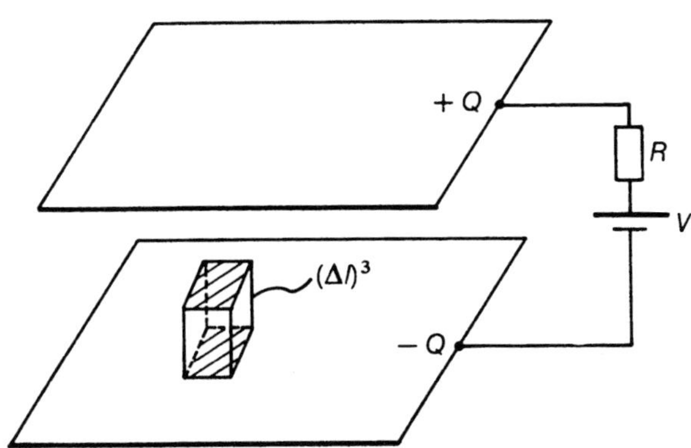

Figure 2.20 The hatched areas define a small 'capacitor' of volume $(\Delta l)^3$.

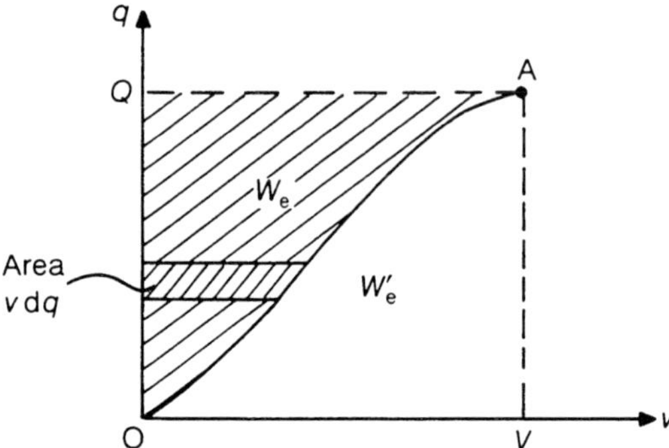

Figure 2.21 Energy W_e associated with the charging of an electric system.

where V is the potential difference between the plates at the end of the charging process.

The charging of an electric field system may be followed in a q–v diagram such as that shown in Figure 2.21. During charging the system is brought from the initial state O ($q = 0, v = 0$), to the final state A ($q = Q, v = V$). The curve OA is, for most common systems, a straight line with a slope equal to the capacitance of the system. Such systems are called linear, to distinguish them from non-linear systems where the capacitance changes during the charging process. The amount of energy stored is $W_e = \int v\,dq$ and is represented by the hatched area between OA and the q axis. The area between OA and the v axis is called the co-energy W'_e. It is always true that $W'_e = QV - W_e$. For the important case of linear systems $W'_e = W_e = 0.5QV$. There is no special significance attached to co-energy, but it is a useful quantity in calculations.

The expression above gives the energy stored in a system of charges. An alternative view is to postulate that this energy is stored in the volume occupied by the field.

Consider a small cube of side Δl, as shown in Figure 2.20. The top and bottom surfaces, shown shaded, are equipotentials, and hence they may be regarded as the plates of a small capacitor. Indeed, they may be replaced by thin conducting sheets without disturbing the field. The energy stored in this cube is then given by $0.5C'V'^2$, where C', V' are the capacitance and potential difference for this small capacitor. Substituting, energy stored $= \frac{1}{2}\varepsilon[(\Delta l)^2/\Delta l](E\Delta l)^2 = \frac{1}{2}\varepsilon E^2(\Delta l)^3$ where E is the electric field between the shaded areas. The energy stored per unit volume is then obtained by dividing by $(\Delta l)^3$:

$$w_e = \tfrac{1}{2}\varepsilon E^2. \qquad (\mathrm{J\,m^{-3}}). \tag{2.24}$$

This is a fundamental formula, generally applicable, and it allows the calculation of the energy storage density anywhere inside a dielectric.

It is useful to consider whether any limit exists to the amount of energy that can be stored in any given volume. Certainly, a high ε would be an advantage. However, the range of physical values of ε is rather limited. The best option is to increase E. There is, however, for every insulating material, a value of the electric field which, if exceeded, results in loss of insulating properties or to what is known as electrical breakdown. For air, under normal conditions, $E_{max} = 30\,kV\,cm^{-1}$. In an engineering system much lower working values must be accepted. Hence for air,

$$w_{e,max} = \tfrac{1}{2}8.85 \times 10^{-12} \times (3 \times 10^6)^2 \simeq 40\,J\,m^{-3}.$$

Although higher values may be achieved using different materials, this is a useful order of magnitude calculation to assess the usefulness of electric field systems for energy storage. It is instructive to contrast this value of energy density with the energy stored, in a much smaller volume, in a car battery (~ 2 MJ), or in 1 kg of fuel oil (~ 40 MJ), or the energy stored as heat in 1 l of water raised from zero to 100 °C (~ 0.5 MJ). The energy density in an electric field is very low. A fuller discussion of the implications of this will be given after magnetic field systems have been discussed.

EXAMPLE E14

The parallel-plate capacitor shown in Figure E14 is insulated by two dielectric layers. The thin air layer has been introduced accidentally during manufacture. Air breaks down at $30\,kV\,cm^{-1}$, and the solid dieletric at $300\,kV\,cm^{-1}$. Calculate the maximum operating voltage for this capacitor so that breakdown is avoided anywhere inside the insulation. Calculate the maximum energy that can be stored per m^2 of capacitor area. Recalculate the above for the case when the entire insulation consists of the solid dielectric.

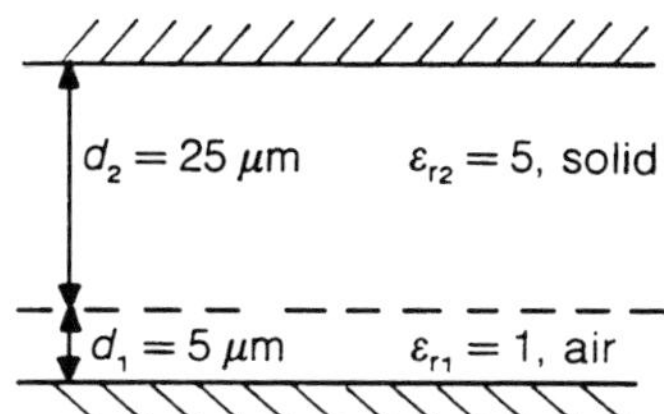

Figure E14 Capacitor with a mixed dielectric.

Solution

Let the electric fields in dielectrics 1 and 2 be E_1 and E_2 respectively. The boundary condition across the dielectric interface is: $D_{n1} = D_{n2}$ or $\varepsilon_{r1}\varepsilon_0 E_1 = \varepsilon_{r2}\varepsilon_0 E_2$, hence $E_1 = 5\,E_2$. Since the maximum permissible field in 1 also has the lowest value, it follows that breakdown in air will limit the maximum operating voltage. Hence, $E_1 = E_{max} = 3 \times 10^6\,V\,m^{-1}$ and $E_2 = \frac{3}{5} \times 10^6\,V\,m^{-1}$. Then $V_{max} = E_1 d_1 + E_2 d_2 = 3 \times 10^6 \times 5 \times 10^{-6} + \frac{3}{5} \times 10^6 \times 25 \times 10^{-6} = 30\,V$, equally shared between the two layers.

$$\begin{aligned}\text{Energy stored per m}^2 &= \tfrac{1}{2}(C_1 V_1^2 + \tfrac{1}{2}C_2 V_2^2)/S^2 \\ &= \tfrac{1}{2}[\varepsilon_{r1}\varepsilon_0(S/d_1)V_1^2 + \varepsilon_{r2}\varepsilon_0(S/d_2)V_2^2]/S^2 \\ &= 0.398\,m\,J\,m^{-2}.\end{aligned}$$

If the air layer was not present, then $E_{max} = 30 \times 10^6\ \mathrm{V\,m^{-1}}$, $V_{max} = 900\ \mathrm{V}$, and energy stored per $\mathrm{m^2} = \frac{1}{2}(\varepsilon_{r2}\varepsilon_0/d)V_{max}^2 = 0.597\ \mathrm{J\,m^2}$. The presence of the air layer reduces significantly the operating voltage and the energy storage capabilities of the capacitor. In the manufacture of capacitors great care is taken to exclude air pockets. This is achieved by pumping all gases and filing voids with insulating oils.

PROBLEM

P17 A parallel-plate capacitor is insulated by three ideal dielectric layers as shown in Figure P17. The maximum permissible electric field in each layer is $E_{1,max} = 300\ \mathrm{kV\,cm^{-1}}$, $E_{2,max} = 100\ \mathrm{kV\,cm^{-1}}$, and $E_{3,max} = 10\ \mathrm{kV\,cm^{-1}}$. Calculate the maximum energy per $\mathrm{m^2}$ of electrode area that may be stored in this capacitor without exceeding the maximum permissible electric field in any layer. Neglect fringing effects.
(*Answer:* $45.7\ \mu\mathrm{J\,m^{-2}}$.)

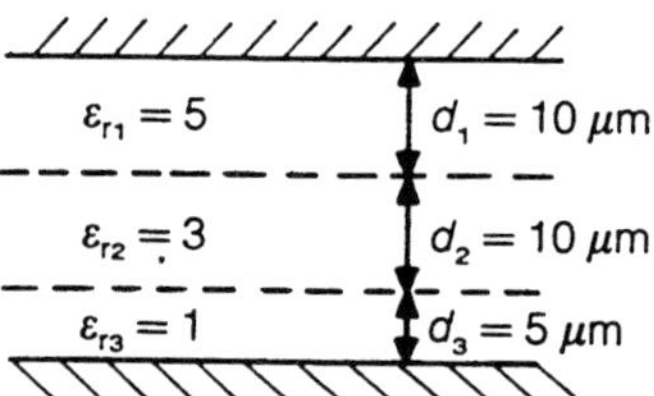

Figure P17

2.8 FORCES IN ELECTRIC FIELD SYSTEMS

Some of the most useful applications of electric fields involve processes where energy is exchanged between the electric field and mechanical systems. It is therefore necessary to study the interaction between mechanical and electrical systems, leading to the development of forces of electrical origin.

Consider the system shown in Figure 2.22 which, although simple, contains all the essential elements of an electromechanical system. The parallel-plate capacitor is connected to an electrical power supply, whilst the moving plate of the capacitor is coupled to the mechanical system. This in turn is modelled by a spring–mass–damper combination. The connection of the moving plate to the electrical supply is via a flexible link to avoid any force exerted through this connection. The problem posed is that of calculating the force of electrical origin $\mathbf{F}_e$ applied to the mechanical system as shown. The position of the moving plate is determined by the distance x, and the convention is adopted that the force $\mathbf{F}_e$ is positive if it acts in the direction tending to increase x (as shown).

Intuitively, it is to be expected that the force will depend on the distance x and the voltage v or, alternatively, on x and the charge q. Similarly, the electrical energy stored will depend on the same variables.

Whatever changes take place in the system, total energy must be conserved, i.e.

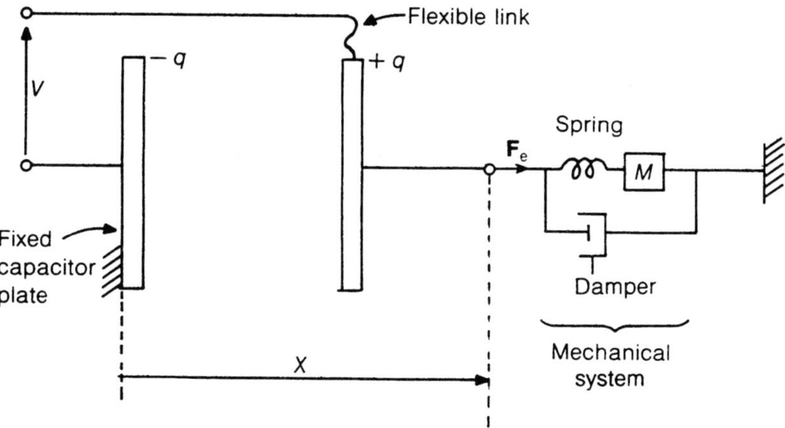

Figure 2.22 The interaction between electrical and mechanical systems.

any increase in the energy stored in the electric field must be equal to the sum of the energies supplied by the electrical source and the mechanical system. If, as a result of the interaction between the electrical and mechanical systems, a charge $\mathrm{d}q$ is transfered and a small displacement $\mathrm{d}x$ takes place, energy conservation demands that:

$$\mathrm{d}W_e = v\,\mathrm{d}q + (-F_e\,\mathrm{d}x). \tag{2.25}$$

The negative sign in the term for the mechanical energy is due to the fact that the force acting on the capacitor plate is equal to $-\mathbf{F}_e$.

Let us now choose x, q as the quantities on which F_e and W_e depend (in mathematical parlance x, q are the independent variables). Then:

$$\mathrm{d}W_e(x, q) = v(x, q)\,\mathrm{d}q - F_e(x, q)\,\mathrm{d}x.$$

This expression must hold for arbitrary values of $\mathrm{d}x$ and $\mathrm{d}q$ (since x, q are independent variables). Hence choosing $\mathrm{d}q = 0$ the following expression is obtained:

$$F_e(x, q) = -\frac{\mathrm{d}W_e(x, q)}{\mathrm{d}x},$$

where the derivative is calculated whilst q is kept constant ($\mathrm{d}q = 0$). Using the notation for partial derivatives introduced before, this becomes

$$F_e(x, q) = -\frac{\partial W_e(x, q)}{\partial x}. \tag{2.26}$$

Similarly, if the potential is desired, it may be obtained from

$$v(x, q) = \frac{\partial W_e(x, q)}{\partial q}. \tag{2.27}$$

It will be seen from equation (2.26) that calculation of the force requires knowledge of the energy stored as function of x, q. In engineering applications it is normally the case that the operating conditions are established by applying known potential differences. It is therefore preferable to obtain expressions for the force where the independent variables are x and v. This can be done as follows.

Equation (2.25) with the new variables is

$$\mathrm{d}W_e(x, v) = v\,\mathrm{d}q(x, v) - F_e(x, v)\,\mathrm{d}x.$$

Also the energy and co-energy are related by the expression

$$W_e(x, v) + W'_e(x, v) = vq. \tag{2.28}$$

Any small changes in energy and co-energy are constrained by (2.28) to satisfy the expression

$$\mathrm{d}W_e(x, v) + \mathrm{d}W'_e(x, v) = v\,\mathrm{d}q(x, v) + q(x, v)\,\mathrm{d}v.$$

Eliminating $\mathrm{d}W_e$ from the last expression (using the energy conservation equation) yields

$$-\mathrm{d}W'_e(x, v) + q(x, v)\,\mathrm{d}v = -F_e(x, v)\,\mathrm{d}x.$$

Following the same argument as before, the force as a function of x and v is obtained from the expression

$$F_e(x, v) = \frac{\partial W'_e(x, v)}{\partial x} \tag{2.29}$$

and similarly, if the charge is desired then

$$q(x, v) = \frac{\partial W'_e(x, v)}{\partial v}. \tag{2.30}$$

It goes without saying that, whether equations (2.26) and (2.27) or equivalently (2.29) and (2.30) are used, the same results are obtained. The choice of equations for any particular problem is a matter of convenience.

It is easy to modify these formulae to include the important case of rotational motion. In this case position is determined by the angular displacement ϑ (in radians) and the equivalent of the force is the torque T_e (in N m). Hence

$$T_e(\vartheta, q) = -\frac{\partial W_e(\vartheta, q)}{\partial \vartheta}$$

or

$$T_e(\vartheta, v) = \frac{\partial W'_e(\vartheta, v)}{\partial \vartheta}. \tag{2.31}$$

EXAMPLE E15

Derive an expression for the force between the plates of the parallel-plate capacitor shown in Figure E15(*a*).

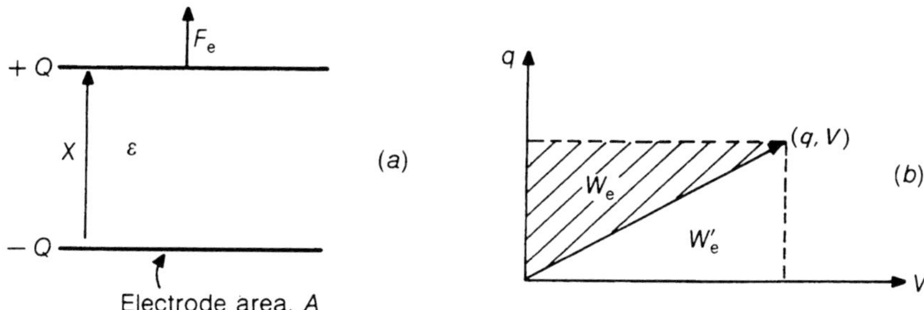

Figure E15 Force between the parallel plates of a capacitor.

Solution

Whichever method is used to tackle this problem, calculation of the electric energy stored is necessary. If the charge and potential difference are q and v respectively, then the energy stored in the capacitor is $W_e = \frac{1}{2}qv = \frac{1}{2}C(x)v^2 = q^2/2C(x)$, where $C(x)$ is the capacitance between the plates equal to $\varepsilon(A/x)$ (edge effects are neglected). The energy stored W_e is shown hatched in the q–v diagram of Figure E15(b). The co-energy is $W'_e = qv - W_e$ and since the system is linear, $W'_e = W_e$ (Figure E15(b)). Using (2.26) and expressing the energy as a function of x and q:

$$W_e(x, q) = \frac{q^2}{2C(x)} = \frac{q^2 x}{2\varepsilon A}.$$

Hence,

$$F_e(x, q) = -\frac{\partial W_e}{\partial x} = -\frac{q^2}{2\varepsilon A}.$$

This expression is always negative and indicates that the force F_e acting on the top plate is in the opposite direction to that shown. The force between the two plates is always attractive.

An alternative approach relies on the use of (2.29). The co-energy as a function of x and v is

$$W'_e = W_e = \tfrac{1}{2}C(x)v^2 = \frac{\varepsilon A}{2x}v^2.$$

Hence,

$$F_e(x, v) = \frac{\partial W'_e}{\partial x} = \tfrac{1}{2}v^2\frac{\mathrm{d}C(x)}{\mathrm{d}x}$$

$$-\frac{v^2}{2\varepsilon A}\left(\frac{\varepsilon A}{x^2}\right)^2 = -\frac{v^2}{2\varepsilon A}C^2(x) = -\frac{q^2}{2\varepsilon A}.$$

As expected, the force is the same as before, irrespective of the choice of mathematical formulation.

The reader may have been tempted to calculate the force using, for example, (2.4).

Such a calculation could proceed as follows:

$$F = -qE = -q\frac{v}{x} = \frac{-q^2}{Cx} = \frac{-q^2}{\varepsilon A}.$$

There is an obvious discrepancy of a factor of two between the two answers. This kind of difficulty arises often when trying to use an expression under circumstances very different to those relevant to its original derivation. Equation (2.4) applies to point charges acted upon by an electric field E. The charge on the top plate of the capacitor is distributed over an area A and, moreover, the electric field is equal to V/x on the side of the charge facing towards the other electrode and zero on the opposite side. Therefore, if one insists on using (2.4), it will be reasonable to use for the electric field an average value $E = (0 + V/x)/2 = V/2x$. The two formulations are then in agreement. In complex systems with distributed charges, it is preferable to use the formulation described in this section.

EXAMPLE E16

Calculate the force acting on a dielectric slab partially inserted between the parallel plates of a capacitor, as shown in Figure E16.

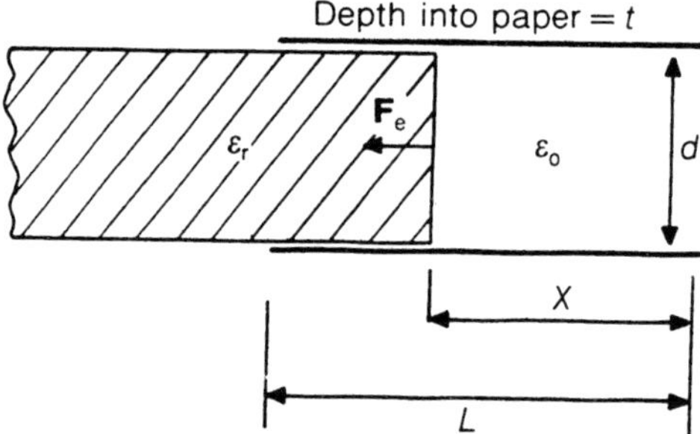

Figure E16 Force on a dielectric insert placed between a parallel plate capacitor.

Solution

The position of the slab is determined by x as shown. The force F_e acting on the slab is by convention positive when it acts in a direction tending to increase x. Since polarization charges will develop in the dielectric slab, it is conceivable that forces will be exerted on these charges. The force may be calculated as in the previous example.

The energy stored in the capacitor is $W_e(x, q) = q^2/2C(x)$, where $C(x)$ is the capacitance due to both the air and slab parts $C(x) = \varepsilon_0(xt/d) + \varepsilon((L - x)t/d)$. In this formulation it is assumed that the slab is sufficiently long and already deep inside the capacitor, so that its small movements do not significantly affect the fringing fields. This means that the correct expression for $C(x)$ differs by a constant term (due to fringing) from that given above. Since in calculating F_e the derivative of $C(x)$ is involved, this correction term does not affect the results. Hence, substituting in (2.26),

$$F_e = \frac{q^2}{2}\frac{C'(x)}{C^2(x)} = \frac{v^2}{2}C'(x) = \frac{v^2}{2}\left(\varepsilon_0\frac{t}{d} - \varepsilon\frac{t}{d}\right) = \frac{v^2 t}{2d}(\varepsilon_0 - \varepsilon).$$

Since $\varepsilon > \varepsilon_0$ the force is negative, i.e. opposite to the direction shown in Figure E16. The slab is pulled in between the plates of the capacitor.

EXAMPLE E17

A small electrode A is placed between a fixed electrode B and can rotate around axis OO′ as shown in Figure E17(*a*). A plan view of the arrangement is shown in Figure E17(*b*). A potential difference V is applied between the two electrodes with the inner electrode kept positive.

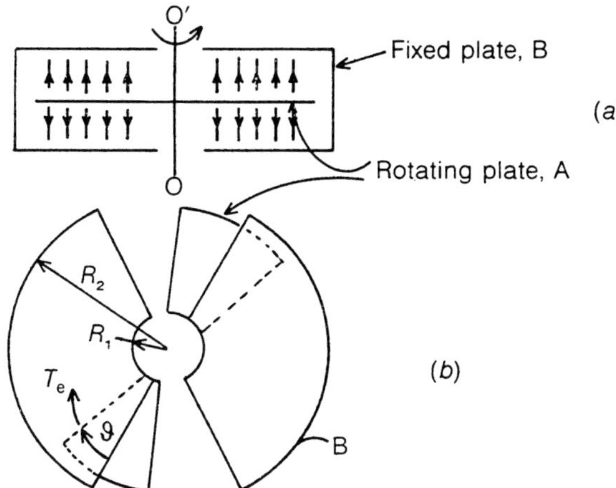

Figure E17 Plate A rotating around OO′ and between the fixed plate B: (*a*) side view, (*b*) plane view.

(*a*) Making simplifying assumptions, obtain an expression for the capacitance between the two electrodes for different values of the overlap angle ϑ.
(*b*) Obtain an expression for the torque acting on the inner electrode.

Solution

There are effectively four capacitors in parallel corresponding to the four field regions shown in Figure E17(*a*). The cross-sectional area of each capacitor is equal to $\pi(R_2^2 - R_1^2)\vartheta/2\pi$. Hence the total capacitance is

$$C(\vartheta) = 4\varepsilon_0(R_2^2 - R_1^2)\vartheta/2d = 2\varepsilon_0(R_2^2 - R_1^2)\vartheta/d.$$

The torque is obtained from (2.31):

$$T_e = \frac{\partial W'_e(v, \vartheta)}{\partial \vartheta} = \frac{\partial}{\partial \vartheta}(\tfrac{1}{2}C(\vartheta)V^2) = \frac{\varepsilon_0(R_2^2 - R_1^2)V^2}{d}.$$

The force is positive, hence it acts in the direction shown and it tends to increase the area of overlap between the two electrodes.

The device described here can be adapted for use as a deflectional instrument to measure potential differences. Instruments of this type are known as 'electrostatic voltmeters' and have a very high input impedance. If a time-varying voltage is applied, the instrument responds to the average value of V^2 and hence it measures the rms value of the voltage.

REMARKS

Examination of equation (2.26) reveals that a force develops if it appears possible that a small displacement dx will lead to a change in the energy stored. If the energy stored does not change, no force or torque can develop. Put another way, the capacitance of the system must be position dependent. In applying equation (2.26) W_e must be expressed in terms of x and q (there must be no explicit dependence on v). Similarly, in (2.29) W'_e must be expressed in terms of x and v (there must be no explicit dependence on q).

PROBLEMS

P18 Two parallel-plate electrodes are immersed in a dielectric fluid and a potential difference V_0 is applied as shown in Figure P18. The mass density of the fluid is ρ_m and the acceleration of gravity is g. Frictional or surface tension forces on the fluid and edge effects are neglected. Derive an expression for the height x to which the fluid will rise between the electrodes.
(*Answer:* $x = 0.5(\varepsilon - \varepsilon_0)(V_0^2/s^2\rho_m g)$.)

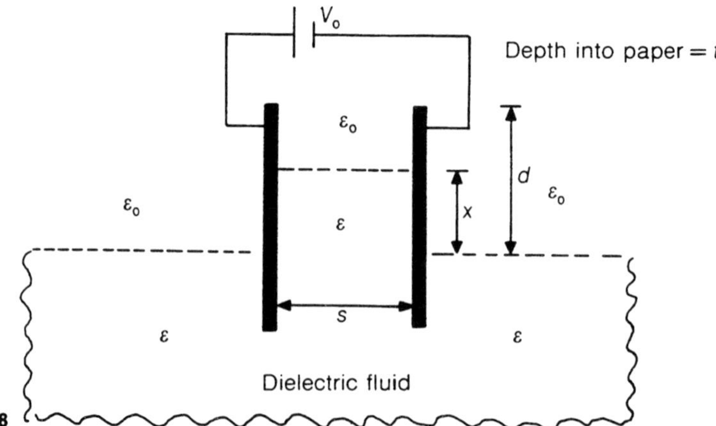

Figure P18

P19 A parallel-plate capacitor with two layers of dielectric is shown in Figure P19. Edge effects are neglected.

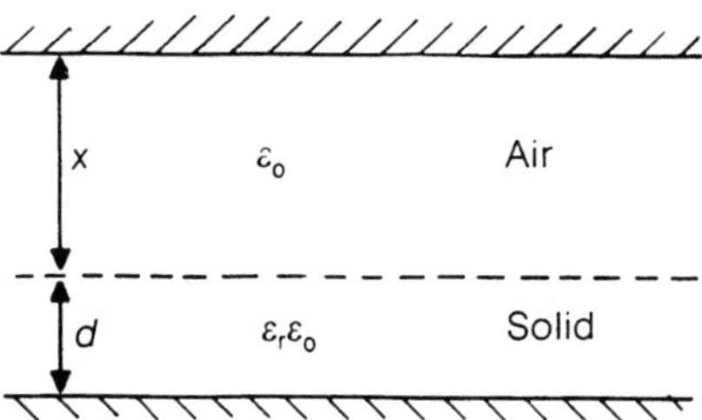

Figure P19

(*a*) Derive a formula for the capacitance between the plates $C(x)$ (for $1\,\text{m}^2$ area on each plate) valid for $x > 0$.
(*b*) Derive a formula for the force acting on the top plate, when a potential difference V_0 is established between the two plates.
(*Answer:* (*a*) $C(x) = \varepsilon_r\varepsilon_0/(\varepsilon_r x + d)$, (*b*) $F_e = -V_0^2\varepsilon_r^2\varepsilon_0/(\varepsilon_r x + d)^2$.)

P20 The two plates of a capacitor are misaligned as shown in Figure P20. Edge effects may be neglected, and the electric field between the two plates may be regarded as uniform in the volume ABCDA′B′C′D′. A potential difference V is applied between the two electrodes. Derive a formula for:

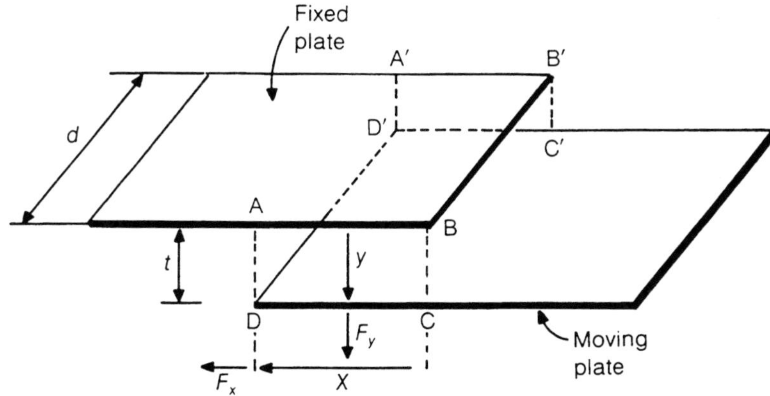

Figure P20 Determination of F_x, F_y between two offset plates.

(*a*) the force acting on the moving plate in the y direction (indicate whether the force is repulsive or attractive);
(*b*) the force acting on the moving plate in the x direction (indicate whether this force tends to align or misalign the two plates).
(*Answers:* (*a*) $-\varepsilon_0 x d v^2/2t^2$ attractive, (*b*) $\varepsilon_0 d v^2/2t$ aligning.)

P21 Two coaxial electrodes are shown in Figure P21. The space between the two

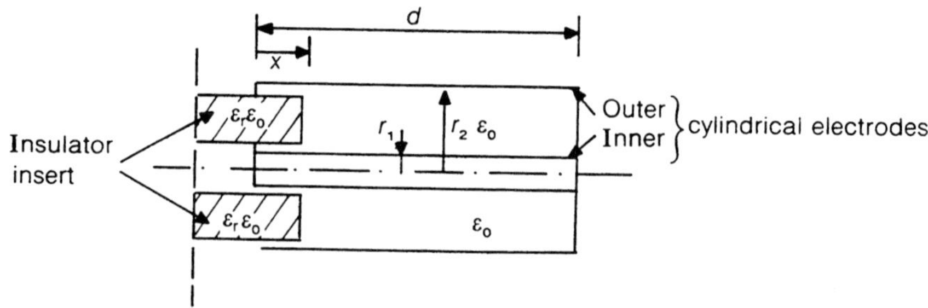

Figure P21 Force on an insulator insert placed inside a coaxial electrode system.

electrodes is filled with air except for a section of length x which is filled by a solid dielectric. Edge effects are neglected.
(*a*) Obtain a formula for the total capacitance $C(x)$ between the two electrodes.
(*b*) A potential difference V_0 is applied between the two electrodes. Obtain an expression for the force acting on the solid dielectric.
(*Answers:* (*a*) $C = 2\pi\varepsilon_0(\varepsilon_r x + d - x)/\ln(r_2/r_1)$, (*b*) $F_e = V_0^2\pi\varepsilon_0(\varepsilon_r - 1)/\ln(r_2/r_1)$.)

CHAPTER 3

Magnetic Field Systems

These systems are typical of situations where electric charges are in steady motion (steady electric current). Charge imbalances are minimal (i.e. there is no charge separation on each conductor) and hence significant electric fields are not present. The type of questions addressed here are: how can forces between current-carrying conductors be calculated; how do different materials behave in the proximity of current carrying conductors; how can an inductor be built and its inductance calculated; are there any limits to the energy that can be stored in an inductor?

3.1 FORCES BETWEEN CURRENT-CARRYING CONDUCTORS

It is an experimental fact that, if a movement of charge (i.e. an electric current) is induced on an otherwise neutral conductor, it sets up an influence around it, such that any other current-carrying coductor in its vicinity experiences a force. Since both conductors are electrically neutral and there is no substantial charge separation, this force cannot be explained in terms of influences set up by static charges (i.e. an electric field). A conceptual model of this interaction between current-carrying conductors can be constructed as follows: each current-carrying conductor sets up around it an influence described as a **magnetic field, B**, such that any other conductor carrying a current I experiences a force proportional to I and B. As an example consider the two long straight conductors carrying currents I_1 and I_2 shown in Figure 3.1. The force experienced by conductor 2 (per metre length), is found to be $F = K_m I_1 I_2 / d$, where K_m is a constant dependent on the magnetic properties of the medium surrounding the conductors and the units. In free space (and approximately in air) and SI units $K_m = \mu_0 / 2\pi$, where μ_0 is the magnetic permeability of vacuum, $\mu_0 = 4\pi \times 10^{-7}\,\mathrm{H\,m^{-1}}$. Hence

$$F = \frac{\mu_0}{2\pi d} I_1 I_2 = \left(\frac{\mu_0}{2\pi d} I_1 \right) I_2.$$

The expression in brackets is the influence of current I_1 on I_2 and is called the magnetic flux density **B** at the position of conductor 2. The magnetic flux density is a vector **B** and is measured in units of tesla (T). (The magnetic flux is measured is units of weber (wb) such that $1\,\mathrm{T} = 1\,\mathrm{Wb\,m^{-2}}$.) For the simple configuration of Figure 3.1,

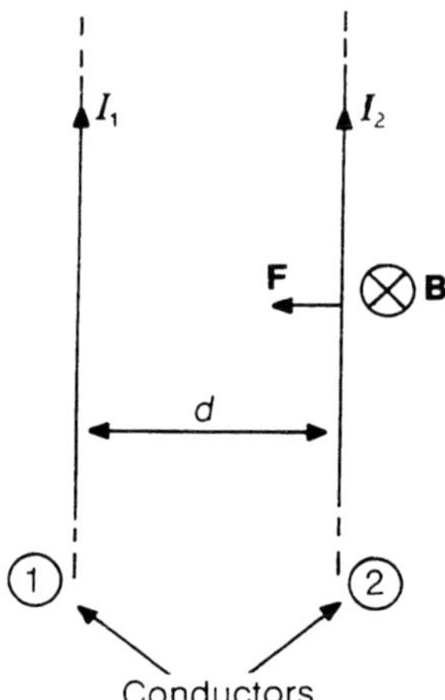

Figure 3.1 Force between two parallel current-carrying conductors.

the value of B at points on conductor 2 is given by the expression $B = \mu_0 I_1/2\pi d$. Experiments using small magnets placed near a long straight conductor show that they also experience a force and that the magnets align in a direction tangential to circles concentric with the conductor as shown in plan view in Figure 3.2. These circles are called the magnetic flux density lines and are analogous to the electric field lines already discussed. The direction of the B lines is found by following the rule shown in Figure 3.3, where, with the thumb of the right hand pointing along the current the fingers indicate the direction of the B-lines. It follows that the magnetic field at any point P, due to a straight current-carrying conductor is perpendicular to the plane defined by the conductor and point P. If the conductor is not straight, then it can be divided into small segments **dl**. Each segment contributes a component to the total field at P given by the Biot–Savart law:

$$\mathbf{dB} = I\mu_0 \frac{(\mathbf{dl} \times \hat{\mathbf{a}})}{4\pi R^2}. \tag{3.1}$$

Here, the vector **dl** has a direction parallel to I and magnitude equal to the length of the small segment. The vector $\hat{\mathbf{a}}$ is directed from the middle of segment **dl** to point P and has a magnitude of unity. R is the distance between the segment and point P. The product $(\mathbf{dl} \times \hat{\mathbf{a}})$ of vectors **dl** and $\hat{\mathbf{a}}$, is known in mathematics as the vector or cross product. The result is itself a vector of magnitude $dl\ a \sin \vartheta$, where ϑ is the angle between vectors **dl** and $\hat{\mathbf{a}}$. The result of this cross product operation is itself a vector perpendicular to both the original vectors **dl** and $\hat{\mathbf{a}}$. Its direction is that in which a right-hand screw advances when the right hand rotates **dl** to coincide with $\hat{\mathbf{a}}$ by the shortest route. The same rule can be used to establish the result of a cross product between any two vectors.

If the magnetic field at a point P due to a conductor of arbitrary shape is required, the conductor is divided into small segments and the field due to each segment is calculated using (3.1). All contributions are then added together (as vectors) to obtain the total value of the field due to the entire conductor. This procedure, although straightforward, can be very tedious for conductors of complex shape.

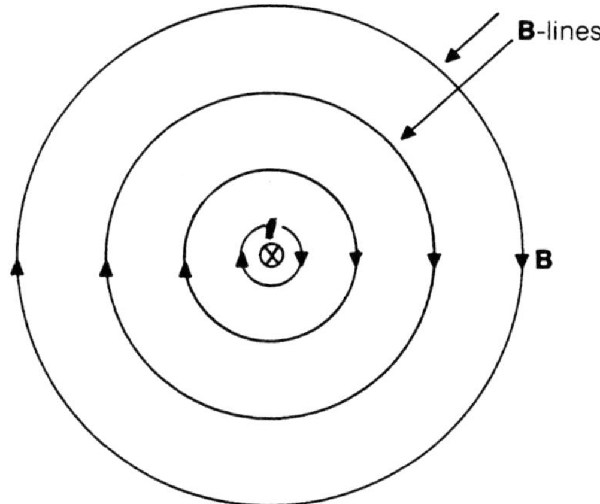

Figure 3.2 Magnetic field lines around a current-carrying conductor.

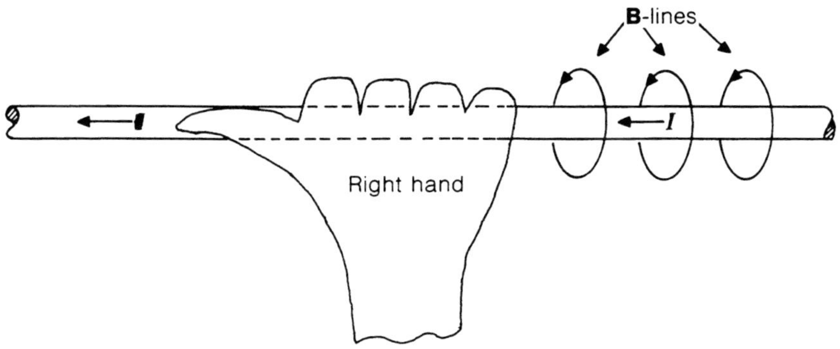

Figure 3.3 Construction for determining magnetic field direction produced by current I.

EXAMPLE E18

Calculate the magnetic flux density at a point a distance d away from a long straight conductor carrying current I (Figure E18).

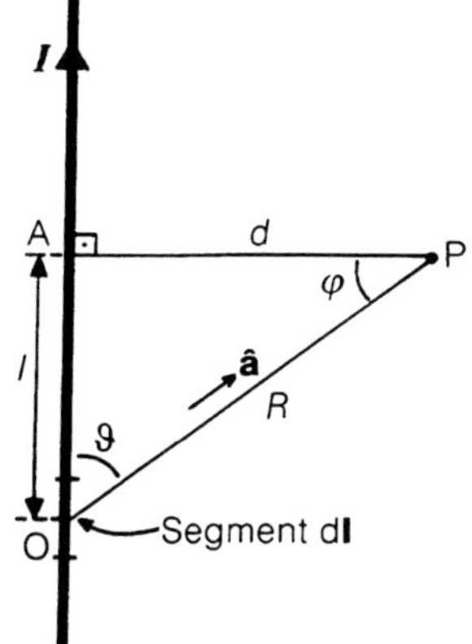

Figure E18

Solution

This problem can be tackled by applying (3.1). A segment dl is chosen around point O on the conductor as shown. The component of the field dB at P due to this segment is parallel to $\mathbf{dl} \times \hat{\mathbf{a}}$ and hence it is directed into the paper. The contributions to the field from all such segments $\mathbf{dl}$ are in the same direction and hence combining them will be easy. However both R and the angle ϑ vary with the position of the segment. This complicates the calculation and it is more convenient if R and ϑ are expressed in terms of d (a constant) and the angle φ. In fact, $\sin\vartheta = \cos\varphi$, $R = d/\cos\varphi$ and $l = d\tan\varphi$. From the last expression $dl = d d(\tan\varphi) = (d/\cos^2\varphi)d\varphi$. Hence, substituting in (3.1),

$$dB = \frac{I\mu_0 \mathbf{de} \times \hat{\mathbf{d}}}{4\pi R^2} = \frac{I\mu_0 dl \sin\vartheta}{4\pi R^2} = \frac{\mu_0 I}{4\pi}\frac{d}{\cos^2\varphi}\frac{\cos^3\varphi}{d^2}d\varphi = \frac{\mu_0 I}{4\pi d}\cos\varphi\, d\varphi.$$

Therefore the total field due to the whole length of the conductor is

$$B = \int dB = \frac{\mu_0 I}{4\pi d}\int_{\varphi=-\pi/2}^{\pi/2}\cos\varphi\, d\varphi = \frac{\mu_0 I}{2\pi d}.$$

$\mathbf{B}$ is pointing into the paper as indicated by the vector product in (3.1) and confirmed by the right-hand rule.

Once the magnetic flux density $\mathbf{B}$ at a point P has been calculated (by using (3.1) for example), the force $\mathbf{dF}$ acting on a conductor element $\mathbf{dl}$ placed at P and carrying a current I can be calculated from the expression

$$\mathbf{dF} = I(\mathbf{dl} \times \mathbf{B}) \tag{3.2}$$

where the vector $\mathbf{dl}$ is taken parallel to I.

The magnitude of the force is $I\,dl\,B\sin\vartheta$, where ϑ is the angle between the conductor and B. The direction of the force is found by using the cross product rules discussed in connection with (3.1). The total force acting on a conductor of arbitrary shape and length is found by combining the elements of force acting on small segments comprising the entire length of the conductor (equation (3.2)).

Let us now relate the force acting on a current-carrying conductor to the force acting on the individual free charges responsible for the current. A small segment dl of a conductor is shown enlarged in Figure 3.4. It is assumed that the current is due to the flow of free electrons in the conductor and, as normal, it is defined as the charge per second traversing a conductor cross-section. By convention, the positive current direction shown is associated with electron flow in the opposite direction. Assuming that there are N free electrons per m^3 in the conductor, drifting on average with a velocity $\mathbf{V}$, then, during a short time interval dt, all free electrons as far as a distance $V\,dt$ away from a cross-section will traverse it. There are $(V\,dt\,A)N$ such electrons. Hence the current is, by definition, $I = dq/dt = (V\,dtA)Ne/dt$, where e is the electronic charge equal to 1.6×10^{-19} C. The current density is then $j = I/A = NeV$ (in $A\,m^{-2}$). The velocity at which electrons drift is proportional to the electric field, with the constant of proportionality known as the electron mobility, μ. Hence, $j = Ne\mu E = \sigma E$, where σ is the electrical conductivity of the

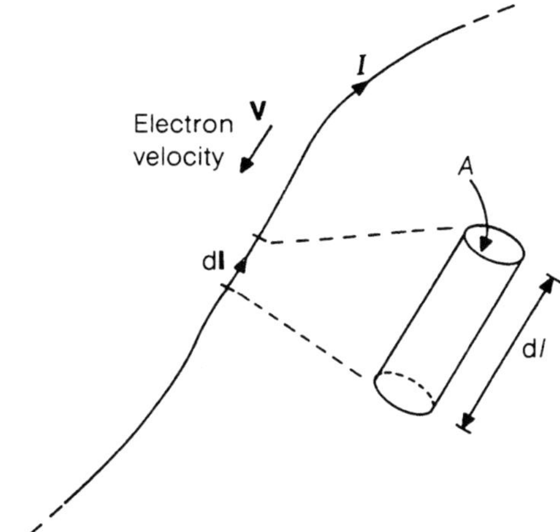

Figure 3.4 Current-carrying conductor and detail.

conductor material. In vector form,

$$\mathbf{j} = \sigma \mathbf{E}. \tag{3.3}$$

Other charges contributing to the current may be treated in the same way. The force acting on the segment in Figure 3.4 under the influence of a magnetic field **B** is then

$$d\mathbf{F} = I(d\mathbf{l} \times \mathbf{B}) = (NeVA)(d\mathbf{l} \times \mathbf{B}) = (-NeA\, dl)\mathbf{V} \times \mathbf{B}$$

since for negative charges **V** and **dl** are antiparallel. The last expression in brackets is equal to the charge q experiencing the force, hence in general

$$\mathbf{F} = q(\mathbf{V} \times \mathbf{B}) \tag{3.4}$$

where q is either positive or negative depending on the polarity of the charge. This formula can be used to give the force acting on a charge q moving with velocity **V** in a region where the magnetic field is **B**. Combining (2.4) and (3.4) gives the general expression for the force acting on a charge q moving with velocity **V** due to the influence of an electric field **E** and magnetic field **B**:

$$\mathbf{F} = q(\mathbf{E} + \mathbf{V} \times \mathbf{B}). \tag{3.5}$$

The first term, $q\mathbf{E}$, gives the force due to the electric field (Coulomb force) and the second, $q\mathbf{V} \times \mathbf{B}$, the force due to the magnetic field (Lorentz force).

REMARKS

Equation (3.1) offers a means by which the magnetic field can be calculated. An alternative method for calculating the magnetic field will be introduced in the next section. It is emphasized again that in every calculation the direction of the field must be ascertained. This can be easily done after the cross product operation between vectors is understood. For calculating the force on conductors (3.2) may be used. Note that the popular expression $F = BIl$ is but a very special case of (3.2), namely the case when the direction of the current and **B** are at right angles to each other. In cases when current conduction takes place over extended regions (not necessarily in thin conductors) it is sometimes useful to calculate the force per unit volume. From (3.2)

$$d\bar{\mathbf{F}} = I\,d\mathbf{l} \times \mathbf{B} = \mathbf{j}\,dA\,d\mathbf{l} \times \mathbf{B} = (dA\,dl)\mathbf{j} \times \mathbf{B}.$$

Hence the force per unit volume is:

$$\mathbf{F} = \mathbf{j} \times \mathbf{B}. \tag{3.6}$$

It is interesting to contrast the nature of the two force terms in (3.5). The energy per second imparted on a material body moving with velocity **V** and acted upon by a force **F** is equal to **F·V**. For the case of an electric charge q in a combined electric and magnetic field, power input $= q(\mathbf{EV} + \mathbf{V}(\mathbf{V} \times \mathbf{B}))$. The first term is non-zero provided that **V** is not perpendicular to **E**. In constrast, the second term is always zero (since $\mathbf{V} \times \mathbf{B}$ is perpendicular to **V** and when the dot product with **V** is calculated it will always result to zero). It follows therefore that a charged particle can exchange energy with an electric field but it cannot do so with a magnetic field. The force acting on a charged particle in a magnetic field is always perpendicular to its velocity. The only motion possible under these circumstances is circular motion around the magnetic field lines. An electron beam can be accelerated by an electric field and collimated by a magnetic field.

PROBLEMS

P22 (*a*) A conductor segment is shown in Figure P22(*a*). Following a procedure

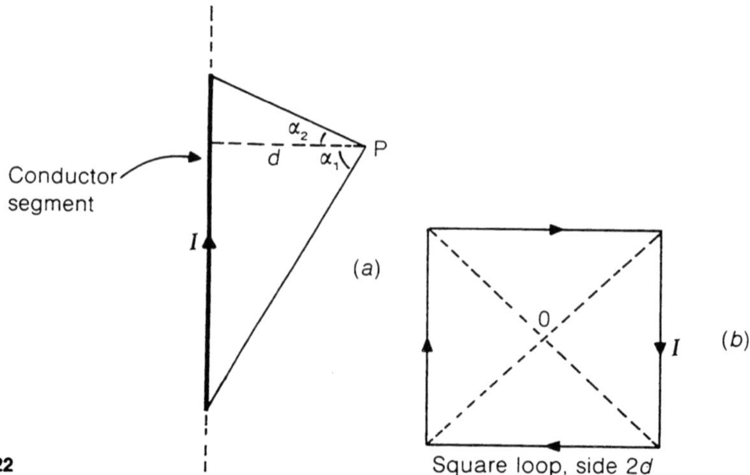

Figure P22

similar to that adopted in Example E18, obtain an expression for the magnetic field at point P.
(*b*) Obtain an expression for the magnetic field at the centre 0 of the conductor loop shown in Figure P22(*b*).
(*Answer:* (*a*) $B = (\mu_0 I/4\pi d)(\sin\alpha_1 + \sin\alpha_2)$, (*b*) $B = 5.65(\mu_0 I/4\pi d)$.

P23 A square loop of wire carrying current I, is placed on the x–y plane as shown in Figure P23. A magnetic field B parallel to the y-axis is applied. Calculate the force (magnitude and direction) acting on the various segments of the loop.
(*Answer:* on sections BC and DA the force is zero, on remaining sections it is equal to BIl resulting in anticlockwise torque.)

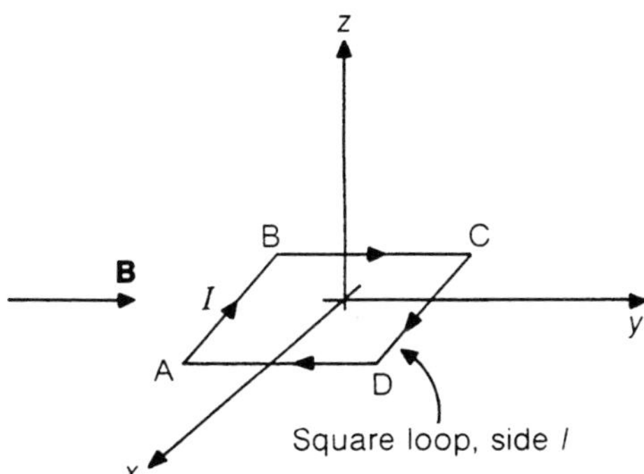

Figure P23

P24 An electron travels with velocity V along the x-axis, when it encounters a magnetic field region where **B** points along the z-axis. Show that the electron trajectory will be circular and derive expressions for its radius r and for the angular frequency of rotation ω.
Hint: at equilibrium the magnetic force is the centripetal force $m_e V^2/r$, where m_e is the mass of the electron. For circular motion $V = \omega r$. ω and r are called the Larmor frequency and radius respectively. Try out some numerical calculations for these quantities.
(*Answer:* $r = m_e V/eB, \omega = eB/m_e$.)

3.2 THE RELATIONSHIP OF THE MAGNETIC FIELD TO ITS SOURCES—AMPERE'S LAW

In the last section a procedure was established for calculating the magnetic field due to small current elements, and it was shown how these contributions can be combined to obtain the field due to a conductor of any shape. In this section an alternative procedure is described which quantifies in an elegant way the relationship of the magnetic field to its sources.

Consider a conductor carrying a current I into the paper as shown in Figure 3.5(a). A closed curve C of arbitrary shape is chosen, as shown. A small

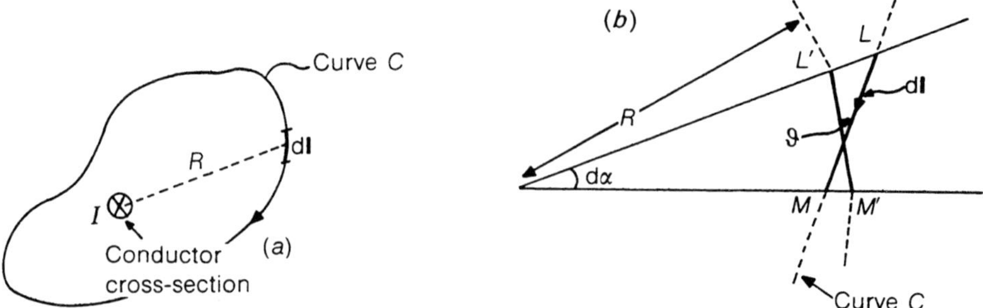

Figure 3.5 Development of Ampere's law.

segment dl of C is also shown expanded in Figure 3.5(b). Its magnitude is equal to the length of the segment, it is tangential to C, and points in the same sense as the magnetic field produced by the current I. The angle dα, defined by the conductor and the extremities L and M of the segment dl, is chosen very small, so that both LM and the arc length L′M′ on the circle centred on the conductor (radius R) can be regarded, approximately, as straight lines. Then, $L'M' = R\,\mathrm{d}\alpha = \mathrm{d}l \cos\vartheta$. Hence the dot product $\mathbf{B}\cdot\mathbf{dl}$, where $\mathbf{B}$ is the flux density at the segment, is equal to $B\,\mathrm{d}l\,\cos\vartheta = BR\,\mathrm{d}\alpha$. For a straight conductor, $B = \mu_0 I/2\pi R$ and therefore $\mathbf{B}\cdot\mathbf{dl} = \mu_0 I\,\mathrm{d}\alpha/2\pi$. This operation is repeated for all segments constituting the entire curve C and all the contributions are added together to form, in effect, the line integral $\mathbf{B}\cdot\mathbf{dl}$ along the closed curve C.

$$\int_C \mathbf{B}\cdot\mathbf{dl} = \frac{\mu_0 I}{2\pi}\int_C \mathrm{d}\alpha = \mu_0 I.$$

If more conductors are present this calculation can be extended to include their contribution. Hence, in general

$$\int_C \mathbf{B}\cdot\mathbf{dl} = \mu_0 \times (\text{total current linked with the curve } C). \tag{3.7}$$

This expression is known as Ampere's law. Care should be taken when using (3.7) to identify the current linked with the curve C. A conductor is 'linked' with C if its removal necessitates cutting the curve C. The 'total current linked', is the algebraic sum of the current in the conductors linked with C. Current appears with a positive sign if it generates a magnetic field in the same direction as that chosen for **dl**. Otherwise, it appears on the right-hand side of (3.7) with a negative sign.

EXAMPLE E19

Obtain a formula for the magnetic field due to a long straight conductor carrying a current I.

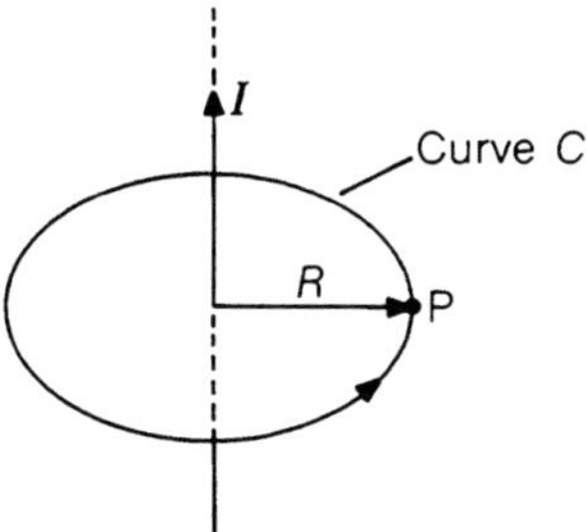

Figure E19 Determination of the magnetic field around a long straight conductor using Ampere's law.

Solution

This is a result already known, but it is instructive to obtain it using Ampere's law. Let us try to calculate the magnetic field at a point P a distance R away from the conductor shown in Figure E19. A closed curve C is chosen, with convenience in mind, exploiting the symmetry of the problem. For the situation depicted in Figure E19, a circle is chosen passing from P and centred on the point where the conductor pierces the plane normal to it and passing from P. On the perimeter of this circle $\mathrm{d}l$ is chosen anticlockwise. From symmetry, B is tangential to C and hence parallel to $\mathrm{d}l$ and also has a constant magnitude on C. Hence,

$$\int_C \mathbf{B}\cdot\mathrm{d}\mathbf{l} = B\int_C \mathrm{d}l = B2\pi R.$$

The total current enclosed is I, and since $\mathrm{d}l$ has been chosen in the direction of the field produced by I, it appears on the right-hand side of (3.7) with a positive sign. Therefore the magnetic field a distance R away is equal, as expected, to $B = \mu_0 I/2\pi R$.

PROBLEM

P25 Four long wire conductors are shown in Figure P25. Ampere's law is applied on each of the three curves C_1, C_2, C_3, shown. In each case obtain the total current linked with each curve.
(*Answer:* $I_2 - I_1, -I_3 - I_4, -I_1 + I_2 - I_3 + 2I_4$.)

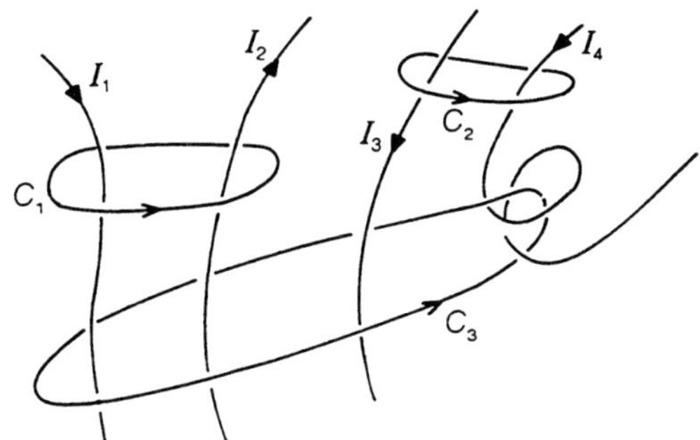

Figure P25

3.3 MAGNETIC MATERIALS

The magnetic field described in previous sections was set up by currents in free space. In engineering practice, however, this is not usually the case. Magnetic fields are set up inside, or in the proximity, of materials other than air, such as, for example, iron. It is therefore necessary to examine how the ideas presented so far can be modified to account for the presence of materials. Not all materials significantly affect the magnetic field. Many common insulating materials, although significantly affecting the electric field, have a negligible effect on the magnetic field. An important class of engineering materials, known as ferromagnetic, affect substantially the magnetic field. Examples of such materials are iron and steel. Other classes of magnetic materials are the diamagnetic and paramagnetic materials. The discussion that follows focuses on ferromagnetic materials as they are the most important in engineering practice. A simplified conceptual model of these materials is formed by regarding them as a collection of a large number of 'elementary current loops' which, in the absence of an applied field, are randomly distributed. Hence their overall effect is to cancel each other out. When an external field is applied, partial or complete reorientation of the elementary loops takes place. It is this effect that determines the magnetic behaviour of the material. It is worth pointing out that the term 'elementary current loops' masks a very complex physical situation. Ferromagnetism has its origins in electron spin interaction (a quantum mechanical effect) which is not elaborated further in this text. With this proviso, let us now study how a magnetic field may be established using a current-carrying coil in air and in a ferromagnetic material.

Consider the situation depicted in Figure 3.6(a), where a very long coil consisting of N turns per metre length is surrounded by air. The current in the coil is I_f, where the subscript f indicates current due to free electrons in the wire. The coil is very long and, from symmetry, it is obvious that the magnetic flux density inside the coil will be in the axial direction as shown. Its magnitude can be calculated by applying Ampere's law on the curve C shown. The length l is arbitrarily chosen, and the segment CD can be placed as far as desired from the axis of the coil. If CD is placed an infinite distance away, the value of **B** along CD will be zero (if this were not the case, an infinite amount of energy would be required to set up the field). Under these

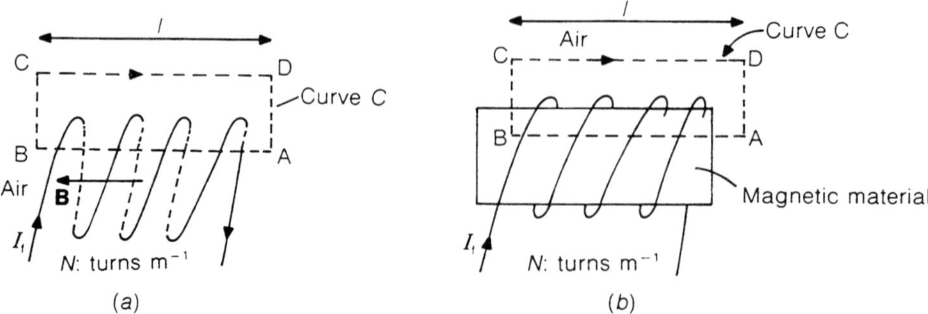

Figure 3.6 Coil (a) without and (b) with ferromagnetic insert.

circumstances, the contribution of CD to the integral will be zero. The contribution of segments BC and DA is more difficult to ascertain. Due, however, to the symmetry of the problem, the field along both segments will be the same. Any contribution made by BC will be exactly cancelled out by DA. Therefore, $\int_C \mathbf{B}\cdot d\mathbf{l} = \int_{AB} \mathbf{B}\cdot d\mathbf{l} = Bl$. Ampere's law thus yields

$$Bl = \mu_0(NlI_f) \qquad \text{i.e.} \qquad B = \mu_0(NI_f).$$

Note, that the current linked is equal to the number of turns linked with C (i.e. Nl) times the current through each turn. The flux density B depends on the medium (μ_0) and on terms independent of the medium (NI_f). It is convenient to define a quantity describing the magnetic field, which is independent of the medium. This quantity is called the magnetic field intensity H (measured in A m^{-1}) and in this example it is equal to NI_f. $\mathbf{H}$ is a vector parallel to $\mathbf{B}$ and in air $\mathbf{B} = \mu_0\mathbf{H}$.

Let us now study the changes in the magnetic field when a cylindrical piece of ferromagnetic material is inserted into the coil as shown in Figure 3.6(*b*). The elementary current loops in the material which hitherto were distributed at random, will now orientate as shown in the cross-sectional view in Figure 3.7(*a*). Internal current elements effectively cancel out and, overall, it appears that a circular 'magnetization current' I_b (in (A m^{-1}) is established as shown in Figure 3.7(*b*). Applying Ampere's law on ABCDA as before

$$Bl = \mu_0(NlI_f + lI_b).$$

Note that the magnetization I_b assists the coil current I_f. Hence

$$B = \mu_0(NI_f + I_b).$$

This expression shows that in order to establish the same flux density as for the coil in air, a smaller coil current is required, as the process is now assisted by the magnetization of the material. It is this property that gives ferromagnetic materials their industrial importance.

Substituting $H = NI_f$, and making the reasonable assumption that the magnetization is proportional to B gives,

$$B = \mu_0 H + x_B(B/\mu_0)$$

where the constant of proportionality x_B is a property of the medium known as the

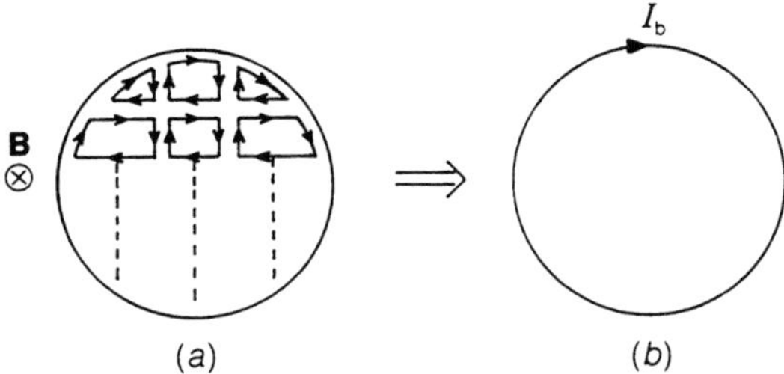

Figure 3.7 Equivalent current I_b describing influence of ferromagnetic material.

magnetic susceptibility. Hence,

$$B = \frac{\mu_0}{1 - \chi_B} H = \mu_r \mu_0 H = \mu H$$

where μ_r is called the relative magnetic permeability of the medium. In most media **B** and **H** are parallel to each other and

$$\mathbf{B} = \mu_r \mu_0 \mathbf{H} = \mu \mathbf{H}. \tag{3.8}$$

Equation (3.8) relates **B** and **H** in any material. Every expression derived so far is valid in the presence of materials, provided that μ_0 is replaced by $\mu = \mu_r \mu_0$. In particular, Ampere's law is more conveniently expressed as follows:

$$\int_C \mathbf{H} \cdot \mathbf{dl} = \text{total free current linked with } C. \tag{3.9}$$

Note that if (3.7) is used instead of (3.9) the total current linked includes both free current (i.e. current in conductors) and also magnetization current (I_b). Equation (3.9) is the preferred form of Ampere's law although, naturally, it is entirely proper to use (3.7).

REMARKS

It is instructive at this stage to contrast the properties and fundamental laws relevant to electric and magnetic fields. Electric field lines start from positive charges and terminate on negative charges. As a result the total electric flux through a closed surface is equal to the total charge enclosed (Gauss's law, $\int_S \mathbf{D} \cdot \mathbf{ds} = Q$). In contrast, magnetic field lines are closed lines—they do not start or terminate anywhere. This is a consequence of the fact that there are no isolated 'magnetic charges' and, hence, the total magnetic flux through a closed surface is always equal to zero ($\Phi = \int_S \mathbf{B} \cdot \mathbf{ds} = 0$, magnetic flux conservation). In a static electric field $\int_C \mathbf{E} \cdot \mathbf{dl}$ along a closed curve is always equal to zero. In contrast, in a magnetic field, $\int_C \mathbf{H} \cdot \mathbf{dl}$ along a closed curve is equal to the total current linked with the curve (Ampere's law).

Applications of the techniques developed in this section will be described in section 3.5.

3.4 CONDITIONS AT THE BOUNDARIES BETWEEN DIFFERENT MAGNETIC MATERIALS

A magnetic field established in a part of space consisting of different magnetic materials is subject to constraints on either side of the boundary between these materials (see for comparison the conditions for the electric field obtained in section 2.5). Consider the boundary between two such materials shown in Figure 3.8(*a*). The boundary conditions can be established by exploiting magnetic flux conservation and Ampere's law. The flux crossing surface S shown in

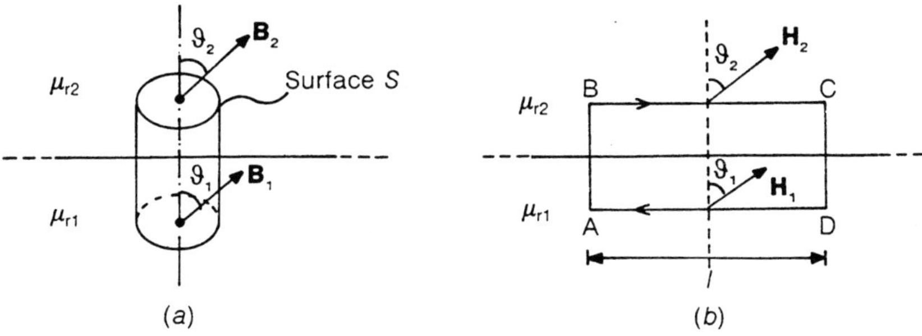

Figure 3.8 Conditions at the boundary between two magnetic materials.

Figure 3.8(*a*) consists of the contributions from the curved and the two flat surfaces. The contribution of the curved surface is zero since the height of the cylinder can be made arbitrarily small. This in turn implies that the conditions thus established only apply very near the boundary. The contribution of the top flat surface is $B_2 \cos \vartheta_2 A$, where A is the area and B_2 the value of the flux density there. The contribution of the bottom flat surface is similarly $-\mathrm{B}_1 \cos \vartheta_1 A$, where the minus sign indicates that the flux component is going into the volume bounded by the surface. If the component B_1 normal to the boundary is designated by B_{n1}, and B_{n2} is similarly defined, then the boundary conditions can be expressed as

$$B_{n1} = B_{n2}. \tag{3.10}$$

A second boundary condition may be obtained by applying Ampere's law on the closed curve ABCDA shown in Figure 3.8(*b*). Since there are no current-carrying conductors at the boundary, the total current linked is zero. Segments AB and CD can be made arbitrarily small and hence they do not contribute to the integral. Segment BC contributes $H_2 \sin \vartheta_2 l$ and segment DA contributes $-H_1 \sin \vartheta_2 l$. The minus sign in the latter case is due to the fact that the component of $\mathbf{H}_1$ parallel to $\mathbf{dl}$ points in the opposite direction to that in which the segment is traversed (D to A). Hence, from (3.9), $H_1 \sin \vartheta_1 = H_2 \sin \vartheta_2$. If the component of H_1 parallel (tangential) to the boundary is designated by H_{t1}, and H_{t2} is similarly defined, then the boundary condition can be expressed as

$$H_{t1} = H_{t2}. \tag{3.11}$$

Equations (3.10) and (3.11) are a complete set of boundary conditions for the magnetic field near the boundary between different materials. A relationship between the angles ϑ_1 and ϑ_2 can be easily established by dividing (3.11) by (3.10), which results after some algebra in $\tan \vartheta_1 / \tan \vartheta_2 = \mu_{r_1}/\mu_{r_2}$.

3.5 CALCULATION OF THE MAGNETIC FIELD IN SIMPLE CONFIGURATIONS—INDUCTANCE

It is relatively easy to calculate the magnetic field established by a simple system of conductors. Consider the very common configuration, known as a two-wire line,

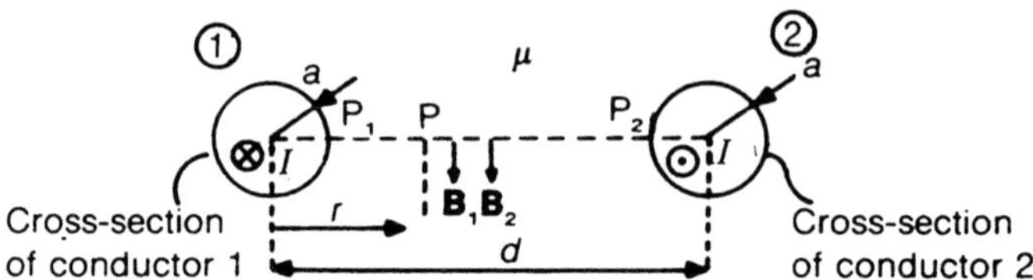

Figure 3.9 Magnetic field due to a two-wire line.

shown in Figure 3.9. It is assumed that the two wires are sufficiently far apart, so that the current distribution in each conductor is unaffected by the presence of the other (i.e. proximity effects are negligible). Moreover, it is assumed that the current flows uniformly along the surface of each wire (a situation common at high frequencies). The magnetic flux density may be calculated everywhere in space by applying superposition. Let us, however, restrict the task to calculating **B** on points along the line P_1P_2. At a point P a distance r away from conductor 1, the magnetic field is the resultant of the influence of conductors 1 and 2. Their contributions are $\mathbf{B}_1$ and $\mathbf{B}_2$ respectively and for the currents shown they are additive:

$$\mathbf{B}(r) = \mathbf{B}_1 + \mathbf{B}_2 = \frac{\mu I}{2\pi r} + \frac{\mu I}{2\pi(d-r)}.$$

The total magnetic flux linked with the two-wire line (i.e. cutting the surface normal to the paper and passing from line P_1P_2) must now be calculated taking account of the fact that $\mathbf{B}(r)$ is strongly dependent on r. Consider the small hatched area ($\mathrm{d}r\, l$) shown in Figure 3.10. The amount of flux crossing this surface is ($B(r)\,\mathrm{d}r\, l$), hence, the total flux linked with the line over a length l can be obtained by adding together all contributions until the whole area $P_1P_2P_3P_4$ is accounted for, i.e.

$$\text{flux} = \int_{r=a}^{d-a} B(r) l\, \mathrm{d}r = \frac{\mu I l}{2\pi}\int_{r=a}^{d-a}\left(\frac{1}{r} + \frac{1}{d-r}\right)\mathrm{d}r = \frac{\mu I l}{\pi}\ln\left(\frac{d-a}{a}\right).$$

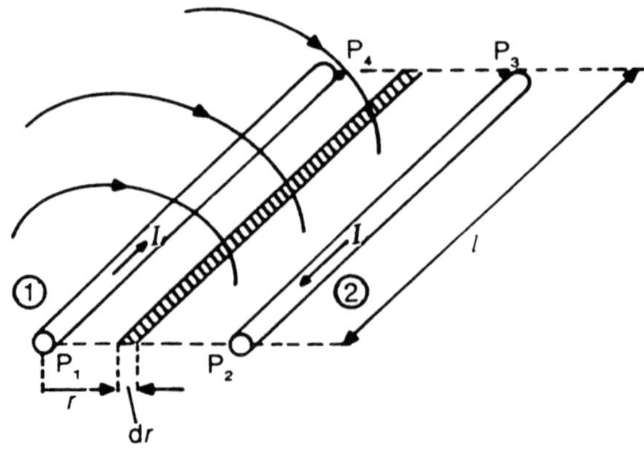

Figure 3.10 Flux linkage for a two-wire line.

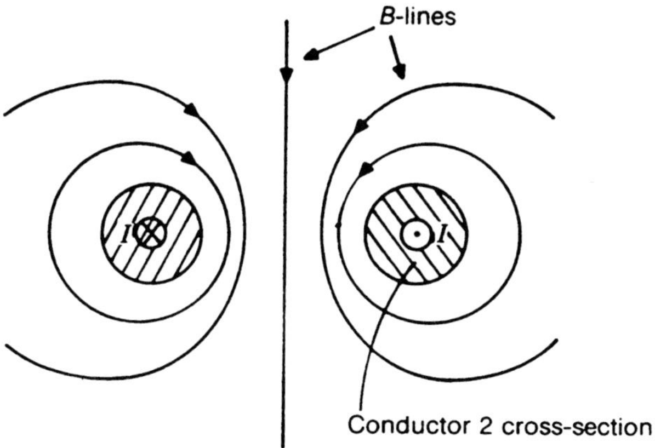

Figure 3.11 Magnetic field lines around a two-wire line.

Hence the flux per unit length linked with the two-wire line is

$$\lambda = \frac{\mu I}{\pi} \ln\left(\frac{d-a}{a}\right) \qquad (\mathrm{Wb\,m^{-1}}).$$

(The flux is measured in Wb; $1\,\mathrm{T} = 1\,\mathrm{Wb\,m^{-2}}$.)

The flux linked with this two-wire line, divided by the current that produced it, is a very important parameter known as the *inductance* L of the line.

$$L = \frac{\text{flux linked}}{\text{current producing this flux}} \tag{3.12}$$

The inductance is measured in henrys (H) and, taking the flux linkage per metre, it follows that the inductance per unit length is

$$\frac{L}{l} = \frac{\mu}{\pi} \ln \frac{d-a}{a} \qquad (\mathrm{H\,m^{-1}})$$

The magnetic field lines on a cross-section of the two-wire line are shown in Figure 3.11.

EXAMPLE E20

Calculate the magnetic field in the coaxial line shown in Figure E20(*a*) and obtain an expression for its inductance per unit length.

Solution

The following general procedure is suggested.

- Ascertain the distribution of current in the conductors, paying particular attention to any obvious symmetries.

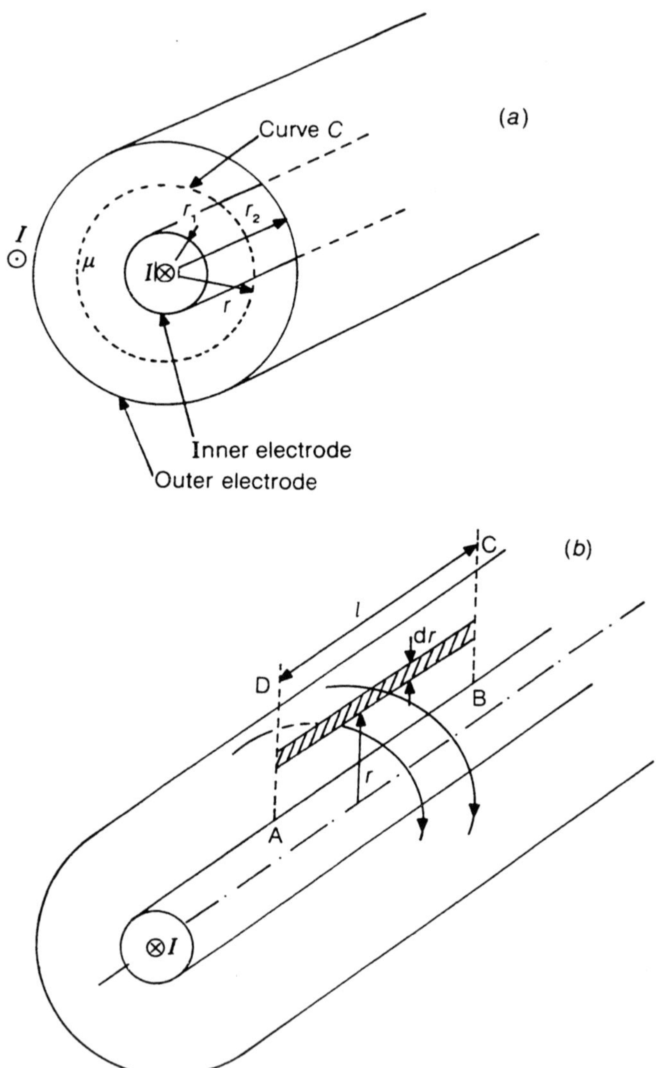

Figure E20 (a) A coaxial line and (b) flux linkage.

- Apply Ampere's law on a closed curve chosen with ease of calculation in mind (exploit any symmetries in the field)
- Calculate the magnetic flux linked with the system of conductors
- Obtain the inductance by dividing the flux linked by the current that produced it

Let us follow this general procedure in the case of the coaxial line.

Assume a current I flowing into the paper and in the centre conductor and returning on the outer conductor. Because of cylindrical symmetry, the current is distributed uniformly on each conductor. It is further assumed that the current in the inner conductor is flowing in its outer surface. Because of symmetry the obvious choice of path to apply Ampere's law is the circular curve, C, shown in Figure E20(*a*). On this curve B is constant in magnitude and always tangential to it. From (3.9), $2\pi r H(r) = I$, since only the current on the inner conductor is linked with C. Hence,

$$B(r) = \mu H(r) = \frac{\mu I}{2\pi r}.$$

Let us now calculate the magnetic flux linked with a length l of the coaxial line. This is the flux crossing the surface ABCDA shown in Figure E20(*b*). A small element of the surface is shown hatched. The flux through it is equal to $(B(r) l \, dr)$. The total flux linked is found by summing up all such contributions for the entire area ABCDA. This is equivalent to integrating from $r = r_1$ to $r = r_2$. Hence, the total flux linked is

$$\phi = \int_{r=r_1}^{r_2} B(r) l \, dr = \frac{\mu I l}{2\pi} \ln \frac{r_2}{r_1} \qquad \text{(Wb)}.$$

The flux per unit length divided by the current gives the inductance per unit length

$$\frac{L}{l} = \frac{\mu}{2\pi} \ln \frac{r_2}{r_1} \qquad (\mathrm{H\,m^{-1}}).$$

Let us now study a class of problems, of great practical importance, where conductors are embedded in media containing a mixture of materials. The most common example is a coil would on an iron former surrounded by air (Figure 3.12). The relative magnetic permeability of iron is assumed to be much larger compared with that for air. Since ferromagnetic materials assist the establishment of magnetic flux, it follows that most of the magnetic flux linked with the coil will be confined inside the iron circuit. A typical B-line is shown in Figure 3.12. A B-line which is not fully contained within the iron circuit is also shown. It represents what is known as the 'leakage flux'. Leakage is usually undesirable, and in a well designed circuit will be small. In many problems leakage may be neglected. A correction for leakage can be made using empirical formulae, or it can be accurately calculated using numerical techniques. Except when otherwise stated, leakage will be neglected in calculations.

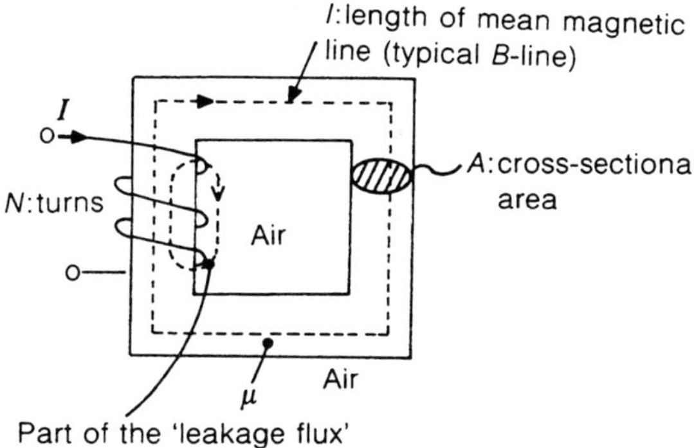

Figure 3.12 Coil wound on iron former.

The cross-sectional area A of the iron is assumed to be constant and since leakage is neglected, flux conservation demands that the product (BA) on any cross-section of the iron remains constant. This in turn implies that B, and hence also H, are constant throughout the iron. Ampere's law applied on a curve coinciding with a typical B-line gives

$$Hl = NI$$

where l is the mean length of a typical B-line.

Therefore

$$\phi = BA = \mu_r \mu_0 HA = \frac{NI}{(l/\mu_r \mu_0 A)}.$$

The denominator, $l/\mu_r \mu_0 A$, is known as the reluctance $\mathscr{R}_m$ of the magnetic circuit. It is a measure of the difficulty with which a flux ϕ can be established in the circuit energized by a current NI (refered to as the magnetomotive force, mmf)

$$\phi = NI/\mathscr{R}_m. \tag{3.13}$$

The reluctance $\mathscr{R}_m$ can be regarded as the equivalent of 'resistance' in an electric circuit. This analogy should not be taken too far, as the resistance is a loss element and the reluctance is not. The flux linked with the coil is $\lambda = N\phi$, hence the coil inductance is

$$L = \frac{\lambda}{I} = \frac{\mu_r \mu_0 N^2 A}{l} \qquad \text{(H)}.$$

A more complicated magnetic circuit is shown in Figure 3.13. As before, the problem can be tackled by demanding flux conservation and by applying Ampere's law. A typical B-line is shown traversing the air gap by the shortest route. Leakage is neglected as before but, obviously, the main flux crosses the air gap. Two questions arise. Firstly, what is the structure of the field in the air-gap region? Two

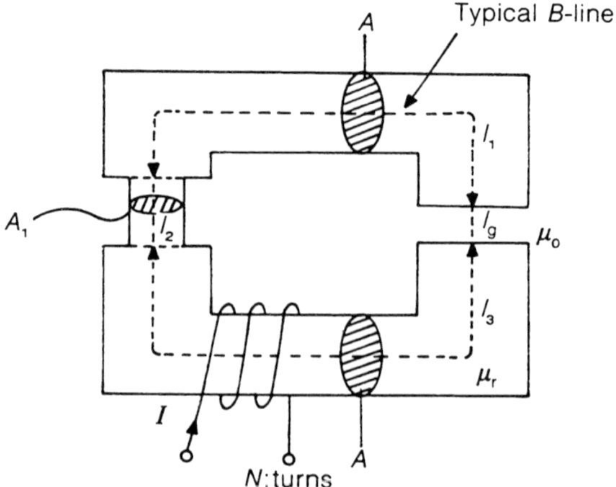

Figure 3.13 A magnetic circuit with non-uniform cross section.

Figure 3.14 Idealization of the magnetic field (*a*) in an air gap and (*b*) fringing.

possibilities are shown in Figure 3.14. In (*a*) the effective cross-section of the gap is the same as in the iron, whilst in (*b*) it is somewhat larger. The situation depicted in (*b*) is known as 'fringing' and it is closer to reality. However, when the gap l_g is small compared with its lateral dimensions, it is reasonable to neglect fringing. The second question concerns the importance of the gap compared with the rest of the magnetic circuit. If the length of a typical B-line in the iron is say 100 mm and in the gap 1 mm, is it permissible to neglect the contribution of the gap in the calculation? The answer to this question is definitely not, and for the following reason. The reluctance of any segment of the magnetic circuit is $\mathscr{R}_m = l/(\mu_r\mu_0)$. Taking typically $\mu_r = 500$ for the iron, l/μ_r is then found to be only 20% of its value in the gap. The reluctance of the gap is five times larger and hence it is the most influential factor in determining the value of the magnetic flux.

Let us now proceed with the calculation of the coil inductance. Leakage and fringing are neglected. Flux conservation demands that

$$B_1 A = B_2 A_1 = B_3 A = B_g A$$
$$\mu_r\mu_0 H_1 A = \mu_r\mu_0 H_2 A_1 = \mu_r\mu_0 H_3 A = \mu_0 H_g A.$$

Hence, $H_3 = H_1 = (A_1/A)H_2$ and $H_g = (\mu_r A_1/A)H_2$. Note that, since $\mu_r \gg 1$, the magnetic field intensity H_g is large in the air gap.

Ampere's Law applied on the typical B-line gives $NI = H_g l_g + H_1 l_1 + H_2 l_2 + H_3 l_3$, and substituting in terms of H_2

$$H_2 = \frac{NI}{\mu_r \frac{A_1}{A} l_g + \frac{A_1}{A} l_1 + l_2 + \frac{A_1}{A} l_3}.$$

Hence

$$\phi = B_2 A_1 = \frac{NI}{\frac{l_g}{\mu_0 A} + \frac{l_1}{\mu_r\mu_0 A} + \frac{l_2}{\mu_r\mu_0 A_1} + \frac{l_3}{\mu_r\mu_0 A}}$$

or $\phi = NI/\mathscr{R}_m$, where the total circuit reluctance is equal to the series combination of reluctances of all circuit segments in the path of the flux.

The inductance of the coil is then

$$L=\frac{\lambda}{I}=\frac{N\phi}{I}=\frac{\mu_r\mu_0A_2N^2}{\mu_r\frac{A_1}{A}l_g+\frac{A_1}{A}l_1+l_2+\frac{A_1}{A}l_3}.$$

The techniques developed so far can deal effectively with many practical situations. It is possible to extend further the range of problems that can be tackled, by using the method of images first described in connection with electric fields. Consider a conductor above a conducting plane as shown in Figure 3.15. The strategy for solving this problem is to ascertain the conditions at the boundary defined by the conducting plane, and then find an image current conductor in air which, together with the original wire conductor, satisfies these boundary conditions. It may be intuitively apparent to the reader, that the magnetic field below the plane conductor will be everywhere zero. To confirm this consider the curve C shown in Figure 3.15, where ABCD extends to infinity. Since there is no current enclosed by this curve, Ampere's law demands that the tangential component of the field along AD is zero. If a non-zero normal component were present then, from symmetry, it would be directed from A to B and from C to D. Ampere's law on curve C (where now ABCD does not extend to infinity) would give a zero contribution from AD (since $H_t=0$) and a non-zero contribution from ABCD (since H would have a component parallel to this path).This, however, cannot be, since the current enclosed is zero. Therefore the inevitable conclusion is that the magnetic field below the conducting plane is zero. Hence, from (3.10), the normal component of **B** at the boundary and above the plane is also zero. The field just above the boundary has a tangential component only. Note that this does not violate (3.11) which was derived on the assumption that there are no free currents at the boundary. This is not the case in Figure 3.15, and the value of H_t at any point on the boundary is equal to the current per unit length flowing on the plane conductor. An image current, flowing in the opposite direction to that of the cylindrical conductor, and placed in air at the same distance and on the other side of the boundary, satisfies the boundary conditions of the original problem. The original conductor and its image can be used to obtain the magnetic field in the configuration of Figure 3.15 (see the derivation in connection with Figure 3.9).

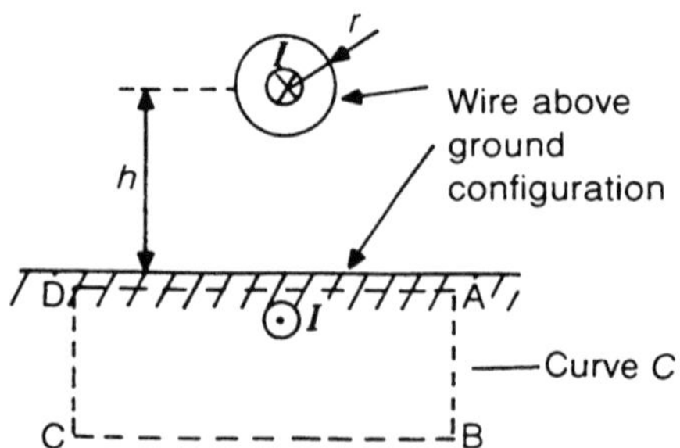

Figure 3.15 Magnetic field for the wire above ground configuration.

EXAMPLE E21

A long wire conductor (current I), is placed in air and parallel to a block of ferromagnetic material of very high permeability ($\mu_r \to \infty$) as shown in Figure E21(a). Find an image conductor suitable for calculating the magnetic field in the air region.

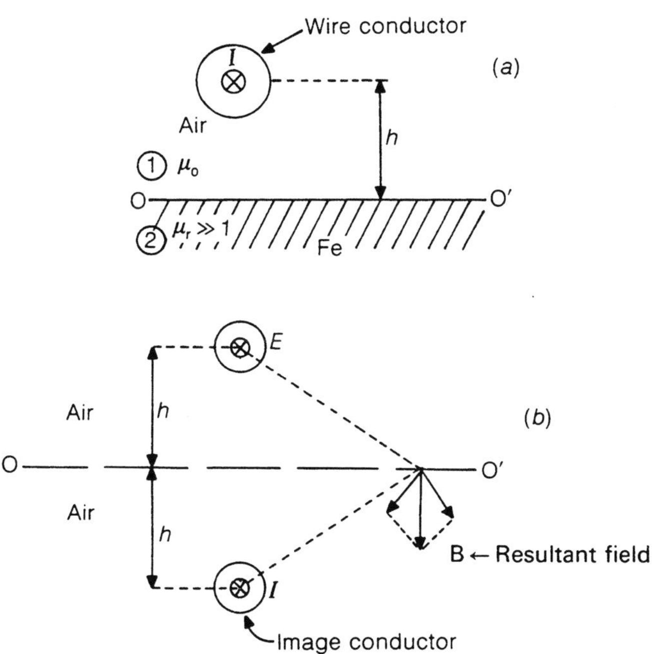

Figure E21 (a) Wire above iron and (b) image.

Solution

At the air/iron boundary, the angles which the magnetic field makes with the vertical in each medium are related by the expression $\tan\vartheta_1/\tan\vartheta_2 = \mu_{r_1}/\mu_{r_2}$ (see section 3.4). In the present problem $\mu_{r_1} = 1$ and $\mu_{r_2} \gg 1$, hence $\tan\vartheta_1 \simeq 0$ and $\vartheta_1 \simeq 0$. This means that on the air side of the boundary the field lines must be normal to it. Consider the conductor configuration shown in Figure E21(b), where an image conductor carrying current I in the same direction as the real conductor has been placed. The field at the boundary due to these two conductors is normal as shown. Hence, the problem depicted in Figure E21(b) satisfies the same boundary conditions in air as the problem in Figure E21(a). The solution to E21(b) above plane OO′ is the same as the solution to E21(a). The image problem is easily solved as both the conductor and its image are in air.

PROBLEMS

P26 A long cylindrical conductor of radius a, carries a current I. Derive formulae for the magnetic field, valid over all space, for the following two cases:
(a) the current flows uniformly on the conductor surface;
(b) the current flows uniformly over the conductor cross-section.
For the case when $I = 10$ A, $a = 2$ mm and uniform current distribution over the conductor surface, plot three B-lines so that the flux per metre between adjacent lines is 2 μWb.
(*Answer*: (*a*) $B(r) = \mu_0 I/2\pi r$ for $r \gg a$, (*b*) $B(r) = \mu_0 Ir/2\pi\alpha^2$ for $r < a$, B-lines at $r = 2$ mm, 5.44 mm, 14.8 mm.)

P27 (*a*) For the problem described in P26(*b*) calculate the value of the inductance associated with the magnetic field inside the conductor.
(b) A rectangular loop of wire is placed near a long straight wire as shown in Figure P27. Derive an expression for the flux ϕ produced by the straight wire and linked with the rectangular loop. Hence find the mutual inductance between the wire and the loop, defined as $M = \phi/I$.
(*Answer:* (*a*) $\mu_0/4\pi$, (*b*) $(\mu_0 b/2\pi) \ln [(d + a)/a]$.

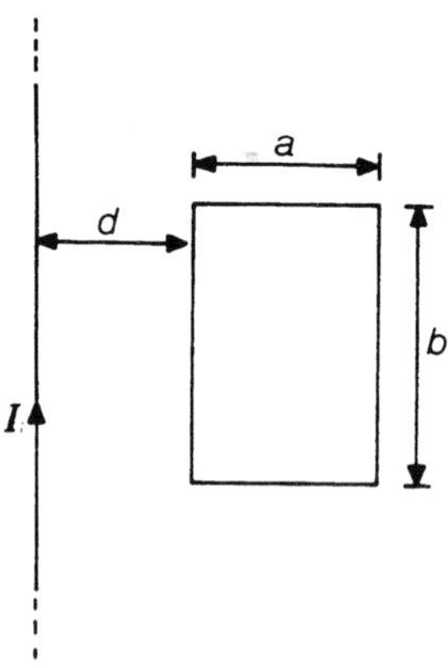

Figure P27 Magnetic coupling between straight conductor and square loop.

P28 A coil of 1000 turns is wound on an Fe former as shown in Figure P28. Calculate the inductance of the coil.
(*Answer:* 0.117 *H*.)

P29 An actuator mechanism is shown in Figure P29. Items 1 and 2 have infinite permeability. All items have the same cross-section S. A typical B-line is shown by the broken line and it is assumed that leakage and fringing are negligible. Derive an expression for the inductance of the coil.
(*Answer:* $L = \mu_0 SN^2/(g + x)$.)

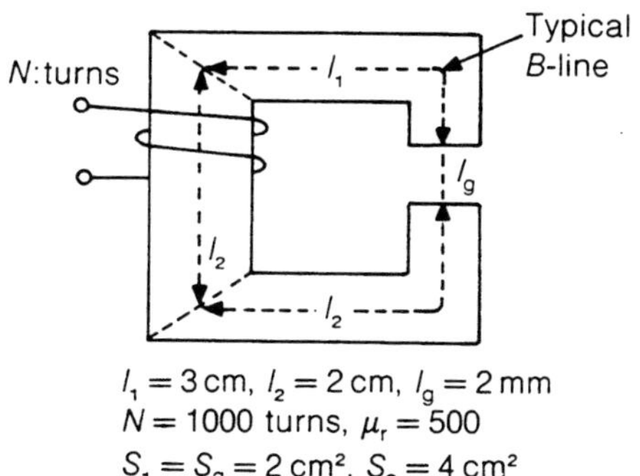

Figure P28

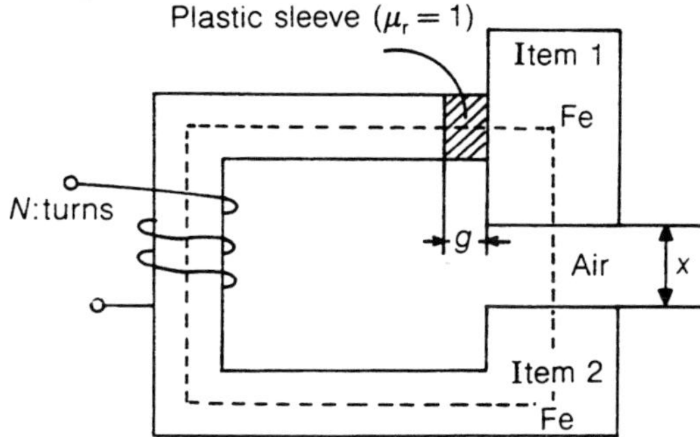

Figure P29

P30 An actuator mechanism is shown in Figure P30. The plunger and the iron yoke have infinite permeability. Obtain an expression for the inductance of the coil. (*Answer:* $L = 2SN^2\mu_0/(x+g)$.)

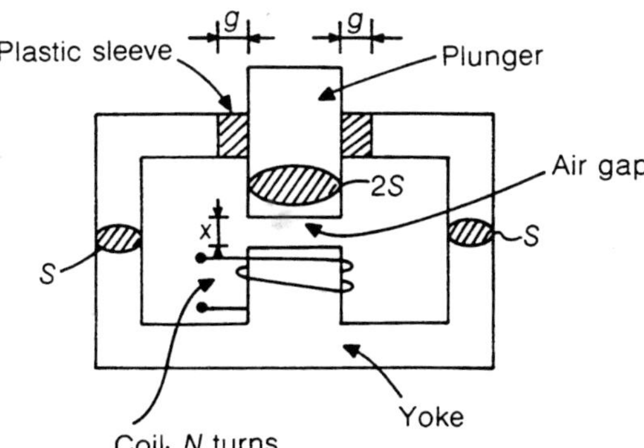

Figure P30

P31 A ring of composite magnetic material is shown in Figure P31(*a*). The material consists of an inner core of low μ, and is surrounded by a layer of high μ as shown in Figure P31(*b*). The mean length of the ring is equal to 2 m. Calculate the ampere–turns (NI) of the coil required to achieve a flux density of 0.2 T in the inner core, and

the total flux through the magnetic circuit. Neglect fringing and assume that the field is constant in each material.
(*Answer:* 637 Ampere–turns, 4.6×10^{-4} Wb.)

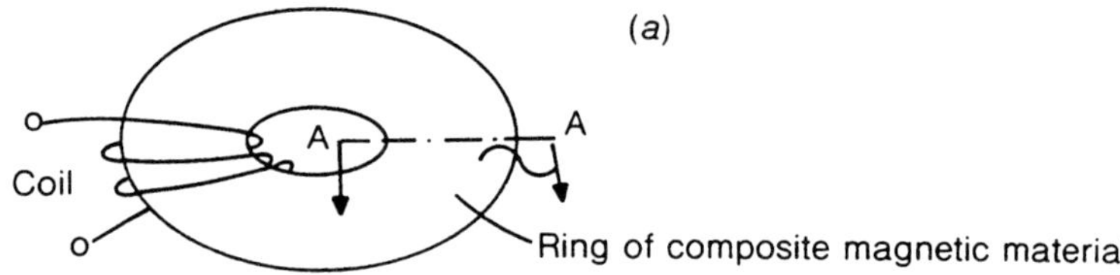

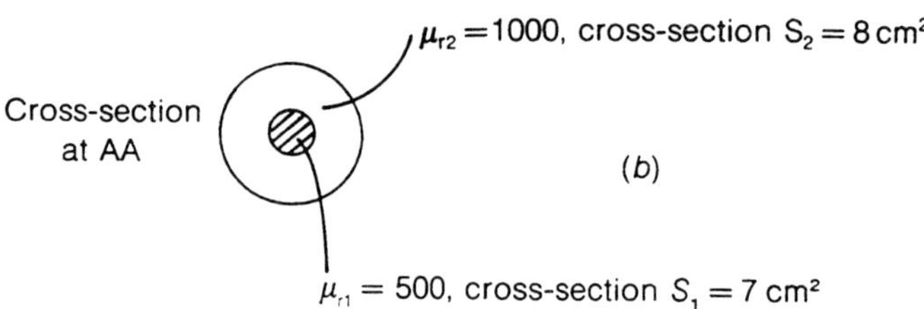

Figure P31 Coil wound on a toroidal former consisting of inner core (μ_{r1}) and outer layer (μ_{r2}).

3.6 FARADAY'S LAW

Up to now it has been assumed that electric fields and potential differences are set up by static charges. It is observed, however, that potential differences are also induced in a circuit when the circuit moves in a magnetic field, or when a magnet moves relative to the circuit, or when there is a changing current near the circuit. The mathematical formulation of these observations is known as Faraday's law and is expressed as follows:

$$V_{ind} = d\lambda/dt \tag{3.14}$$

V_{ind} is the voltage induced in the circuit, and λ is the magnetic flux linked with the circuit. This concept of interdependence between electric and magnetic fields belongs naturally to Chapter 4 as only static phenomena are studied here. However, the establishment of even a static magnetic field is preceded by a time-varying stage, during which energy is supplied to the magnetic field system. It is useful to have a model for this process. Consider the circuit shown in Figure 3.16, where λ is the flux linked with the loop circuit shown. The arrows associated with V, I and λ are the positive reference directions for each quantity. This means that if, for example, the flux λ is in the direction of the arrow, then it is regarded as a positive quantity. As an illustration of the significance of Figure 3.16 and the reference conventions, assume that λ is positive and increasing. Then $d\lambda/dt$ is also positive and hence from (3.14) V_{ind} is also positive. If a resistor is connected across terminals A and B a current will flow from A to B through this resistor. This current in turn will produce a flux in a

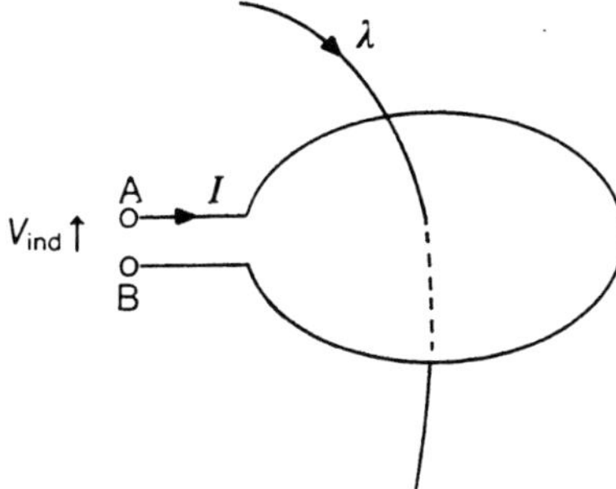

Figure 3.16 Positive reference directions for Faraday's law.

direction opposite to the positive reference flux direction. It will tend to cancel out the original flux increases. This is hardly surprising as, if it were otherwise, any increase in flux would set up a sequence of events leading to further flux increases and hence an instability. This principle is usually stated as Lenz's law, namely that a changing flux induces curents which tend to oppose the change in flux. A fuller study of Faraday's law is postponed until Chapter 4.

PROBLEM

P32 An elementary generator is shown in Figure P32. A rectangular wire loop (dimensions a, b, 100 turns) is rotated with a constant angular frequency $\omega = 314\,\text{rad}\,\text{s}^{-1}$.

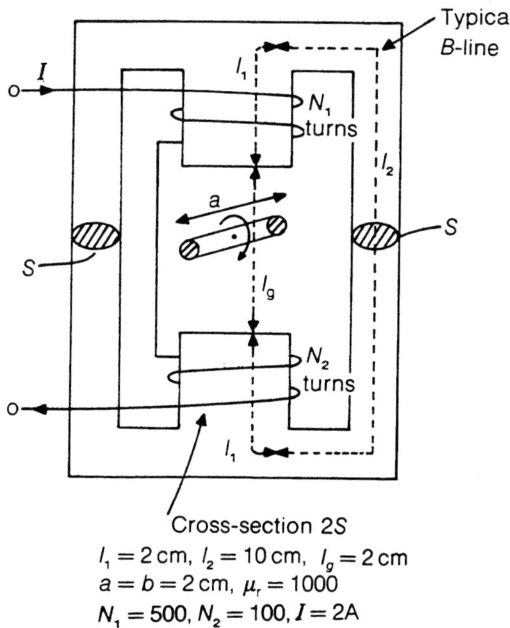

Figure P32

(a) Neglecting fringing and assuming a uniform magnetic field is established in the air gap, calculate the flux density B_g in the gap.
(b) Derive an expression fo the voltage $V(t)$ induced in the loop.
(*Answer:* (*a*) $B_g = 50\,\text{mT}$, (*b*) $V(t) = 0.62\cos(314t)\,\text{V}$.

3.7 ENERGY STORAGE IN A SYSTEM OF CURRENT-CARRYING CONDUCTORS

Many of the most important applications of electromagnetic fields are based on their ability to store energy. The ideas developed here are similar to those explored in section 2.7 in relation to energy storage in an electric field system. Consider as an example the typical magnetic circuit shown in Figure 3.17. Following closure of switch S and a short transient, a current $I_0 = V_s/R_s$ and therefore a flux ϕ_0 will be established as shown. Let us calculate the energy supplied by the source to the circuit on the righι of terminals AA′. Since there are no losses, this energy must be stored in the system of current-carrying conductors. The instantaneous power is $v \times i$ and hence the energy supplied is $\int_0^t v \times i\,dt$. Substituting v from (3.14) results in

$$\text{energy} = \int_0^t \frac{d\lambda}{dt}\, i\,dt = \int_{\lambda=0}^{\lambda_0 = N\phi_0} i\,d_\lambda$$

where λ is the flux linked with the coil. The coil energization process is shown in Figure 3.18. The integral above is the total energy stored and is represented by the

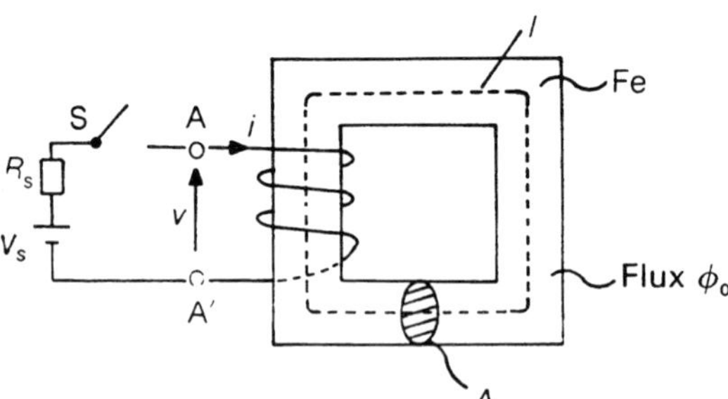

Figure 3.17 Coil energization and establishment of flux ϕ_0.

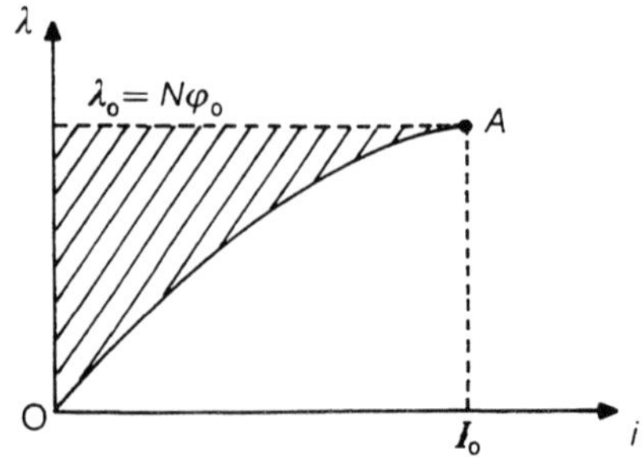

Figure 3.18 The hatched area is a measure of the energy stored in magnetic field system.

hatched area in Figure 3.18. In many systems the path OA is a straight line (linear systems), such that $\lambda = Li$ where L is the inductance of the coil. Hence

$$\text{energy stored} = W_{\mathrm{m}} = \int_0^{\lambda_0} \frac{\lambda}{L}\,\mathrm{d}\lambda = \frac{\lambda_0^2}{2L} = \frac{(LI_0)^2}{2L} = \tfrac{1}{2}LI_0^2.$$

In general, when a current i flows through a coil of inductance L, the energy stored is given by the formula

$$W_{\mathrm{m}} = \tfrac{1}{2}Li^2. \tag{3.15}$$

An alternative view is to postulate that W_{m} is stored in the volume occupied by the magnetic field. The coil inductance was found in section 3.5 to be $L = \mu N^2 A/l$. Hence

$$W_{\mathrm{m}} = \tfrac{1}{2}\frac{\mu N^2 A}{l} I_0^2 = \tfrac{1}{2}\mu Al\left(\frac{NI_0}{l}\right)^2 = \tfrac{1}{2}(Al)H\mu = \tfrac{1}{2}\mu H^2 \times \text{(iron volume)}.$$

Therefore the energy per unit volume stored in the magnetic field is

$$w_{\mathrm{m}} = \tfrac{1}{2}\mu H^2 = B^2/2\mu \qquad (\mathrm{J\,m^{-3}}). \tag{3.16}$$

This is a fundamental formula allowing the calculation of stored magnetic energy per unit volume anywhere inside a magnetic material. It should be compared with the corresponding formula for the electric field energy density (equation (2.24)). As in the case of electric field systems, a magnetic co-energy w'_{m} is defined so that $w_{\mathrm{m}} + w'_{\mathrm{m}} = \lambda i$. For linear systems $w'_{\mathrm{m}} = w_{\mathrm{m}} = 0.5\lambda i = 0.5Li^2$.

REMARKS

Let us consider whether there is a limit to the amount of energy that can be stored in a magnetic field. It is found that in ferromagnetic materials, the flux density is limited by 'magnetic saturation' to values not exceeding 1.5 T (approximately). Higher values are possible but with special materials and a great deal of effort. Hence the maximum energy that can be stored in the air gap of a magnetic circuit is

$$w_{\mathrm{m,\,max}} = \frac{1.5^2}{2 \times 4\pi \times 10^{-7}} \simeq 10^5\,\mathrm{J\,m^{-3}}.$$

This should be contrasted with the corresponding value for electric fields of 40 J m^{-3} (section 2.7). There is a very clear advantage, as far as energy storage is concerned, in using magnetic fields. For the same power rating, a magnetic-type machine would be much smaller compared with an electric-type machine. For this reason, most energy conversion devices rely for their operation on magnetic field interactions.

3.8 FORCES IN MAGNETIC FIELD SYSTEMS

Let us now study the exchange of energy between magnetic fields and mechanical systems, and the forces of magnetic origin thus developed. The treatment of this problem follows closely that of section 2.8 for electric field systems.

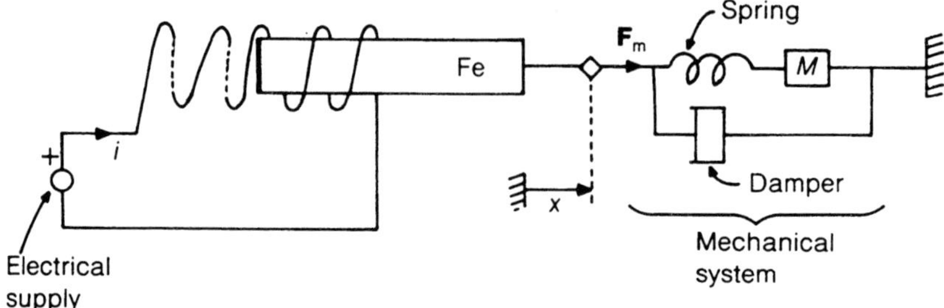

Figure 3.19 Interaction of magnetic field and mechanical systems.

Consider the electromechanical system shown in Figure 3.19. Intuitively it is apparent that, following the establishment of a current through the coil, a force will be exerted on the iron block and hence an interaction with the mechanical system will take place. The objective is to calculate the force of magnetic origin F_m applied to the mechanical system. It is reasonable to suppose that the flux linked with the coil and the force depend on the coil current and the position of the iron block, i.e. $\lambda(i, x)$, $F_m(i, x)$. Energy conservation requires that any change in the magnetic energy stored is accounted for by energy input from the electrical supply and from the mechanical system, namely that

$$dw_m = i d\lambda + (-F_m dx).$$

Since $w_m + w'_m = \lambda i$, it follows that $dw_m + dw'_m = \lambda di + i d\lambda$. Substituting into the energy conservation equation gives

$$\lambda di - dw'_m = -F_m dx.$$

The force is then found to be

$$F_m = \frac{\partial w'_m(i, x)}{\partial x}. \tag{3.17}$$

For rotational motion

$$T = \frac{\partial w'_m(i, \vartheta)}{\partial \vartheta}. \tag{3.18}$$

Note that in linear systems $w'_m = 0.5 Li^2$. Hence (3.17) and (3.18) can be expressed as

$$F_m = 0.5 i^2 (\partial L(i, x)/\partial x) \qquad \text{and} \qquad T_m = 0.5 i^2 (\partial L(i, \vartheta)/\partial \vartheta)$$

respectively. Alternatively, if λ and ϑ are chosen as the independent variables $T_m = -(\partial w_m(\lambda, \vartheta)/\partial \vartheta)$. The reader should study again section 2.8 where a more detailed discussion of similar calculations is given.

EXAMPLE E22

Derive a formula for the force F_m acting on item B in Figure E22. Item A is fixed and both items have zero reluctance. Leakage and fringing are neglected.

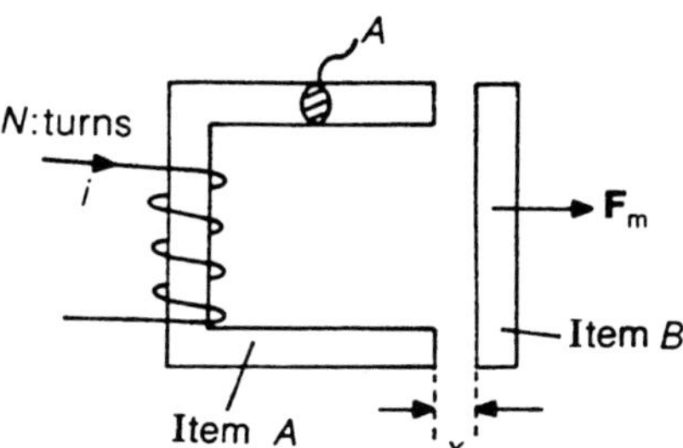

Figure E22

Solution

To work out the force from (3.17) it is necessary to know the co-energy. Flux conservation requires that $B_i = B_g$, hence $H_i = H_g/\mu_r$. Since the iron has zero reluctance ($\mu_r \gg 1$) it follows that $H_i \simeq 0$. Ampere's law on a typical B-line curve gives $Ni = 2xH_g$.

Also $w'_m = w_m = 0.5\lambda i = 0.5(\mu_0(NiAN/2x))i$. The force is then obtained from (3.17) and is found to be $F_m = -\mu_0(Ni)^2A/(4x^2)$. The negative sign indicates that the force is in the opposite direction to that shown, i.e. that it is attractive. Since the force depends on the square of the current, it remains attractive irrespective of the direction of current flow.

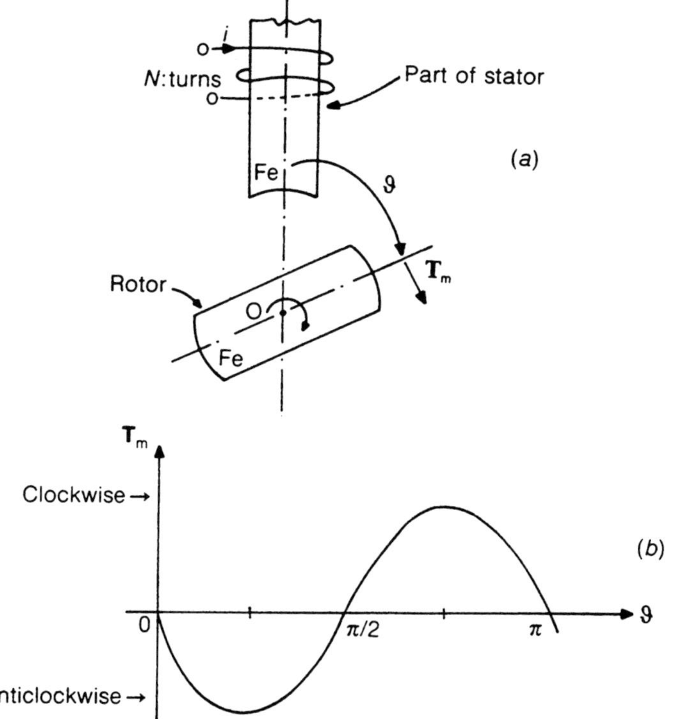

Figure E23 (*a*) Schematic of rotor and stator and (*b*) torque variation with angle.

EXAMPLE E23

Consider the part of a magnetic field system shown in Figure E23(*a*), consisting of a coil would on a stator and a rotor which can rotate around an axis perpendicular to the paper at point O. The inductance of the coil depends on the angle ϑ so that $L(\vartheta) = A + B\sin(\pi/2 - 2\vartheta)$, where A, B are constants. Derive a formula for the torque acting on the rotor.

Solution

The formula for $L(\vartheta)$ is a reasonable one since, as expected, it gives a maximum value for $\vartheta = 0$ and minimum value for $\vartheta = \pi/2$. The co-energy is $w'_m = 0.5[A + B\sin(\pi/2 - 2\vartheta)]i^2$ and hence $T_m = -i^2 B\cos(\pi/2 - 2\vartheta)$. This formula for the torque is plotted in Figure E23(*b*). It is seen that for $0 < \vartheta < \pi/2$ the rotor moves anticlockwise to align with the stator, i.e. it moves to a position of maximum inductance or, equivalently, to a position of minimum reluctance.

EXAMPLE E24

Consider the device shown in Figure E24(*a*) consisting of a stator with six teeth arranged in three groups (phases) a–aa, b–bb and c–cc. Coils are wound in all three phases but to maintain clarity only the one on a–aa is shown. The rotor has no coils but consists of four teeth as shown. Study qualitatively the behaviour of this system as different coils (phases) are energized in turn.

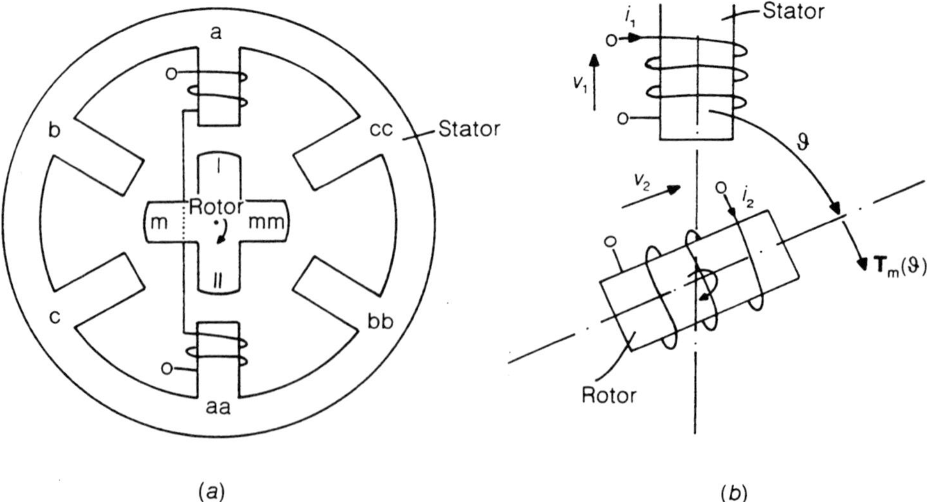

Figure E24 (*a*) Schematic of a stepping motor (only stator is directly excited), and (*b*) a doubly excited system (both stator and rotor coils are excited).

Solution

When coil a–aa is energized, the rotor settles to a position of minimum reluctance (as shown) corresponding to the case when the rotor and stator teeth are in alignment (a–aa aligns with l–ll). Then the current through coil a–aa is switched off and coil b–bb is energized. The position of minimum reluctance is now achieved when b–bb aligns with m–mm. Hence the rotor moves clockwise by an angle $90 - 60 = 30°$. As the coils are energized in the sequence a–aa, b–bb, c–cc, a–aa,... the rotor moves clockwise in steps of 30°. This is an example of a very useful device known as a 'stepping motor'. Note that if the coils are energized in the sequence a–aa, c–cc, b–bb, a–aa,... then the rotor steps in the anticlockwise direction. The speed of rotation is determined by the rate at which current is switched from one coil to another. Consider now the case when with a–aa energized, coil b–bb is also energized. The position of minimum reluctance is neither when a–aa aligns with l–ll nor when b–bb aligns with m–mm. The rotor will in fact settle in a position such that tooth a partially overlaps with tooth 1, and b with m. This corresponds to a step angle of 15°.

Stepping motors are very useful as control and positioning devices as they are easy to control electronically, can be run at low speeds and stalled without damage.

Many practical magnetic systems consist of a rotor and stator, each with a separate coil as shown schematically in Figure E24(*b*). Following the procedure of section 3.7, the energy stored in such a system is

$$W_m = \int_0^t (v_1 i_1 + v_2 i_2)\,dt = \int_0^t \frac{d\lambda_1}{dt} i_1 dt + \int_0^t \frac{d\lambda_2}{dt} i_2 dt$$

$$= \int_0^t d\lambda_1 i_1 + \int_0^t d\lambda_2 i_2$$

where λ_1, λ_2 represent the total flux linked with coils 1 and 2. The flux linked with coil 1 is partly due to current i_1 and partly to i_2:

$$\lambda_1 = L_1 i_1 + M i_2$$

where L_1 is the self-inductance of the coil and M is the mutual inductance of coils 1 and 2 (for a calculation of mutual inductance see, for example, problem P27(*b*) and also Chapter 4). Similarly the flux linked with coil 2 is $\lambda_2 = M i_1 + L_2 i_2$. Hence

$$\int_0^t i_1 d\lambda_1 = \int_0^t i_1 d(L_1 i_1 + M i_2)$$

$$= L_1 \int_0^{I_1} i_1 di_1 + M \int_0^{I_2} i_1 di_2$$

$$= 0.5 L_1 I_1^2 + M \int_0^{I_2} i_1 di_2.$$

Similarly

$$\int_0^t i_2 d\lambda_2 = 0.5 L_2 I_2^2 + M \int_0^{I_1} i_2 di_1.$$

Hence

$$W_m = 0.5L_1I_1^2 + 0.5L_2I_2^2 + M\int d(i_2i_2)$$

or

$$W_m = 0.5L_1I_1^2 + 0.5L_2I_2^2 + MI_1I_2. \tag{3.19}$$

In a linear system $W'_m = W_m$; hence the torque acting on the rotor is

$$T_m = \frac{\partial W'_m}{\partial \vartheta} = 0.5I_1^2\frac{dL_1}{d\vartheta} + 0.5I_2^2\frac{dL_2}{d\vartheta} + I_1I_2\frac{dM}{d\vartheta}. \tag{3.20}$$

These formulae are useful in studying electrical machine behaviour and will be used in section 3.10 on electromechanical energy conversion.

REMARKS

The calculation of force or torque of magnetic origin follows exactly the methods and techniques for force and torque of electric origin. In general more compact and more powerful devices are possible using a magnetic rather than an electric interaction. However, electric-type devices can be designed which may have advantages when low power, high efficiency or fidelity are desired. Examples of magnetic-type devices are many, e.g. actuators, loudspeakers, microphones, motors and generators with ratings up to hundreds of MW. A more detailed look at electromechanical energy conversion will be taken in section 3.10.

PROBLEMS

P33 For the actuator mechanism described in problem P30, derive an expression for the force of magnetic origin acting on the plunger. If the plunger has mass M and is restrained by a spring exerting a force $F_s = K(L' - x)$, where L' and K are constants, obtain the differential equation describing the motion of the plunger. (*Answer:* $F_m = -[N^2\mu_0 I^2/(x+g)^2]S$, $M(d^2x/dt^2) = -[N^2\mu_0 I_S^2/(x+g)^2] - K(L' - x)$.)

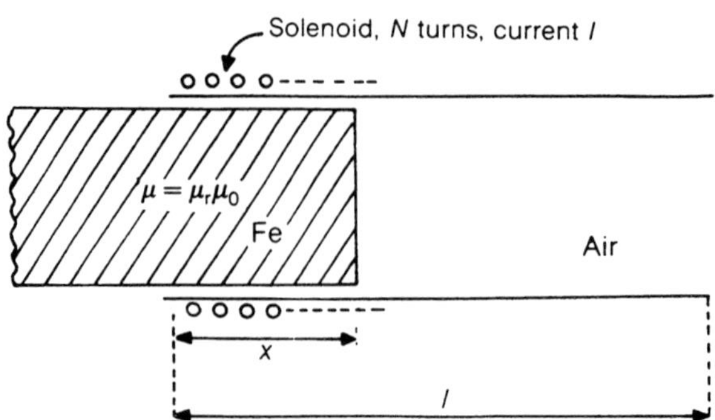

Figure P34

P34 A long plunger of permeability μ is partially inserted into a long solenoid as shown in Figure P34. Find the force exerted on the plunger. (Neglect fringing and assume uniform field distribution on any solenoid cross-section.)
(*Answer:* $F_m = \frac{1}{2}\mu_r\mu_0(\mu_r - 1)[x + \mu_r(l - x)]^{-2}N^2AI^2$.)

P35 A part of a linear actuator is shown in Figure P35. It may be assumed that all the magnetic flux crosses the air gap uniformly over the overlapping tooth sections of the moving and stationary members.

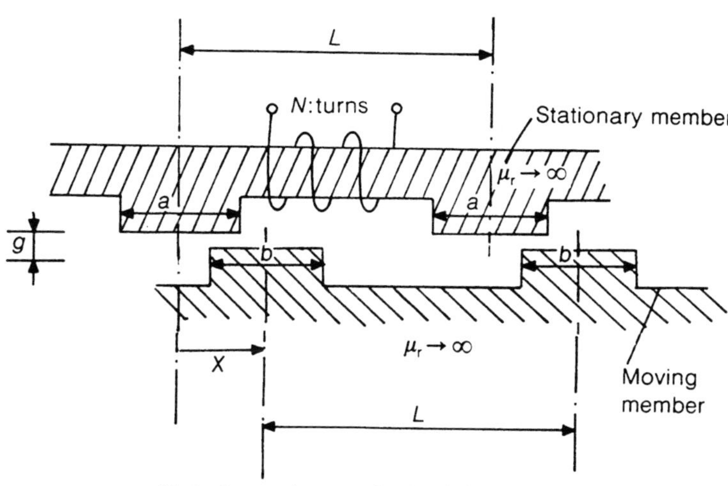

Figure P35 Note: $b < a$, $L > a + b$, depth into paper $= t$

(*a*) Calculate the coil inductance L for $0 < x < L/2$.
(*b*) For $0 < x < L/2$ derive expressions for the force F_x exerted on the moving member, assuming that a constant current I flows through the coil.
(*c*) At which position (approximately) should the current to the coil be switched on to achieve the fastest current build-up and the maximum force over the widest range of x?
(*Answer:* (*a*) $L(x) = (\mu_0 N^2/2g)S(x)$ where $S(x)$ is zero for $x > (a + b)/2$, $S(x) = [(a + b)/2 - x]t$ for $(a - b)/2 < x < (a + b)/2$, and $S(x) = bt$ for $x < (a - b)/2$. (*b*) $F_x = (\mu_0 N^2/4g)I^2(\partial S(x)/\partial x)$. (*c*) Switch on at $x \gtrsim (a + b)/2$.)

P36 An actuator mechanism is shown in Figure P36. Item 1 is fixed, whilst item 2 can slide in the direction shown. The magnetic path through the iron is equal to $l - g$ where l is a constant and g is the gap length. Friction and leakage are neglected.
(*a*) If $\mu_r \to \infty$, derive an expression for the force acting on item 2. What force would be required to pull items 1 and 2 apart when $g = 0$? Disucss your results.
(*b*) Show that for a finite μ_r the force required to separate items 1 and 2 is given by the formula $F_m = (\mu_0 AN^2I^2/2l^2)\mu_r(\mu_r - 1)$.
(*Answer:* (*a*) $F_m = -(\mu_0 N^2I^2A/2x^2$.)

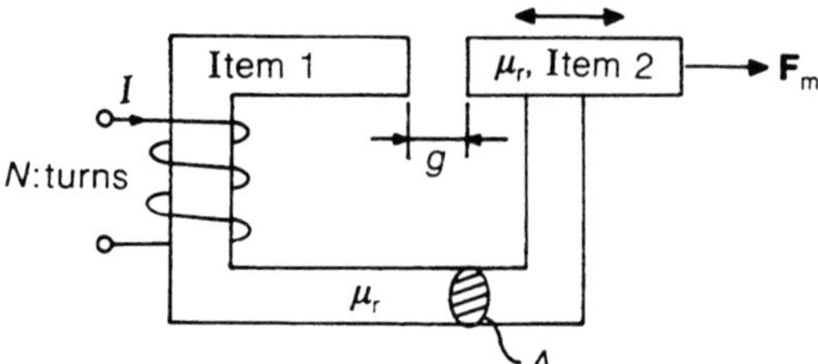

Figure P36

3.9 PROPERTIES OF MAGNETIC MATERIALS—PERMANENT MAGNETS

A phenomenological explanation of magnetization was given in section 3.3. A more realistic model for this process is given in terms of the concept of 'magnetic domains'. In ferromagnets, small segments of material (called 'domains') are magnetized. These domains have small dimensions (typically 100 μm) and their direction of magnetization is random as shown schematically in Figure 3.20. When an external magnetic field is applied, the domains adjust so that their direction of magnetization approaches that of the applied field. A plot of B versus H appears as a straight line (path Oa in Figure 3.21). There is clearly a limit to the process of alignment and, at some stage, further increases in H do not produce proportional increases in B (path aa′ in Figure 3.21). This behaviour is called magnetic saturation and in practice, for

Figure 3.20 Magnetic domains.

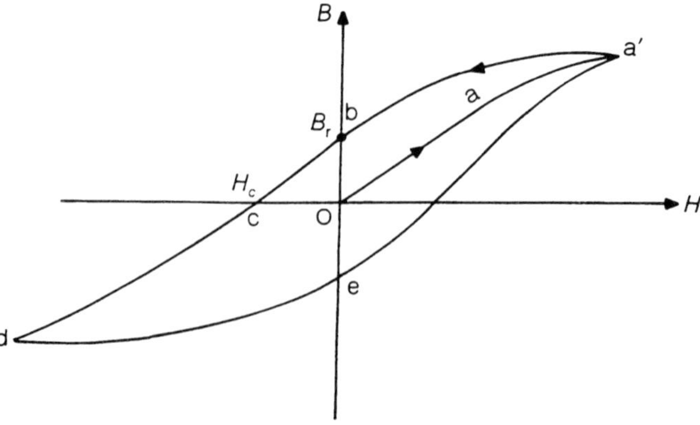

Figure 3.21 B–H curve (hysteresis loop).

most common materials, limits the maximum flux density that can be achieved to approximately 1.5 T. Note that in a typical magnetization process, such as that shown in Figure 3.12, the coil current determines the value of H (Ampere's law). In the saturation region small increases in B require much larger increases in H and, hence, a very high current. This is clearly impractical. Eventually the slope $\Delta B/\Delta H$ of the B versus H curve approaches the value μ_0.

Let us now assume that after the process of magnetization has reached point a′ in Figure 3.21, the coil current, and hence H, are decreased to zero. It is found that the path traced is not a′ao but ab as shown. This phenomenon is known as magnetic hysteresis. It is explained by the fact that, after removal of the current ($H = 0$), the magnetic domains are unlikely to return to their original random state. The value of the flux density at point b is called the remanent flux density B_r. If the coil current changes polarity (H negative), point c is reached where the flux density is zero. The value H_c when this happens is known as the coercivity of the material. Further increases in current bring the state of magnetization to point d. A reduction in current (de) and polarity reversal bring the state of magnetisation along path ea′ to the original point a′. The loop a′bcdea′ is known as the hysteresis loop of the material. It will be shown in section 5.3 that the area of the hysteresis loop is a measure of energy losses in the magnetic circuit.

Consider the magnetic circuit of Figure 3.22(*a*). A current is established in the coil so that the state of magnetization of the material is represented by point a in Figure 3.22(*b*). The coil current is then reduced to zero and the material is brought to point b on the B–H curve. The remainder of the B–H curve is shown by the broken line. Since the coil current is zero, the coil can be removed, but the material will remain at b on the B–H curve. Let us now study what changes will take place when a

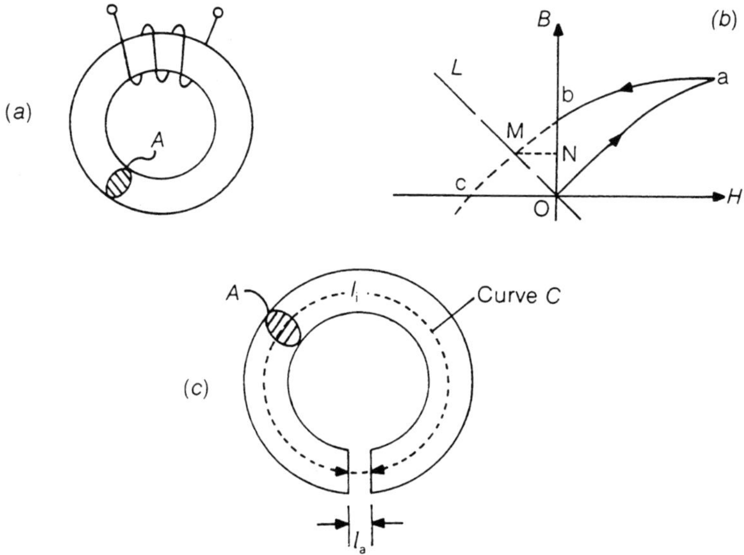

Figure 3.22 Construction and operating region of a permanent magnet.

small slice of thickness l_a is removed from the magnetic circuit as shown in Figure 3.22(*c*). The magnetic field intensity will change say to H_i and H_a in the iron and air gap respectively. From Ampere's law applied on curve *C*, the following expression is obtained $H_i \times l_i + H_a \times l_a = 0$ (since there is no current linked with this curve). From flux conservation and neglecting leakage and fringing, $B_i = B_a = \mu_0 H_a$. Eliminating H_a gives $(-\mu_0 l_i/l_a)H_i = B_i$. This linear relationship between H_i and B_i is ploted in Figure 3.22(*b*) (line OL, slope $\Delta B/\Delta H = -\mu_0 l_i/l_a$). It is now evident that since the state of magnetization of the material must lie on line OL and also on the B–H curve, the only possible solution is that the iron operates at the point M, marking the intersection between OL and the B–H curve. A permanent magnet has thus been constructed with a flux density in air gap equal to ON. A permanent magnet always operates on portion cb of its B–H curve. It is obvious that the exact shape of the curve cb will affect significantly the all-important flux density in the gap. For a material to be suitable for a permanent magnet it is not sufficient that it has a high remanent flux density (high Ob). A better figure of merit can be obtained by seeking to maximize the flux density B_a in the air gap in the arrangement shown in Figure 3.23.

From Ampere's law $B_a l_a/\mu_0 = -H_i l_i$, and from flux conservation $B_a S_a = B_i S_i$. Multiplying these two expressions gives:

$$B_a^2 = -\mu_0 \frac{(l_i S_i)}{(l_a S_a)} (H_i B_i).$$

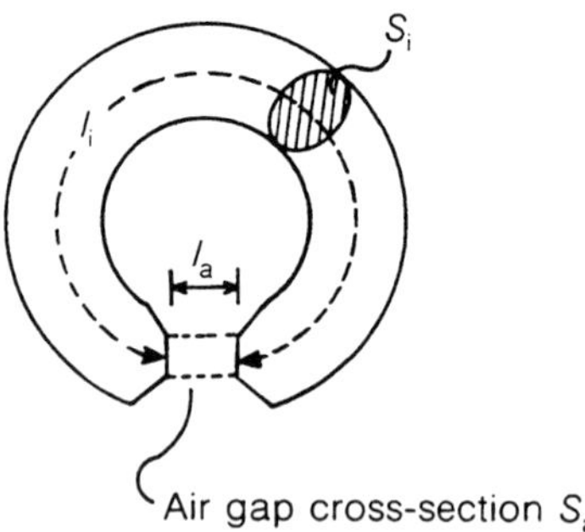

Figure 3.23 Permanent magnet and flux concentration in the air gap.

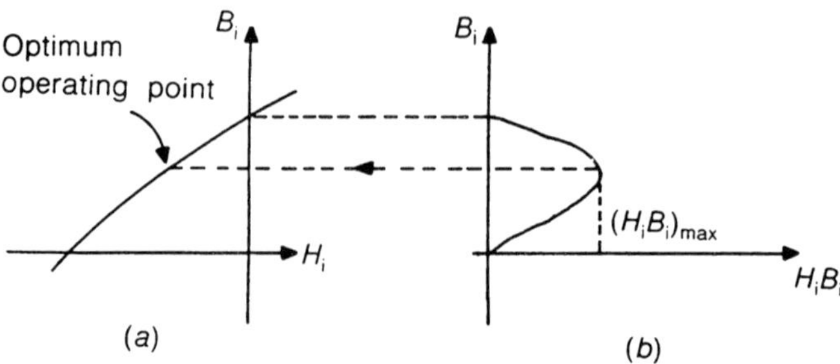

Figure 3.24 Energy product and optimum operating point for permanent magnets.

The maximum value of B_a is achieved when the permanent magnet is operating at a point on the B–H curve making $(H_i \times B_i)$ a maximum. This quantity is therefore a figure of merit for the material, known as the energy product. When comparing two different materials which have the same volume $(l_i S_i)$ with a view to establishing the maximum flux density in an air gap of a given volume $(l_a S_a)$, the best material is the one with the highest maximum energy product $(H_i B_i)_{max}$. The energy product curve for the material described by the B versus H curve of Figure 3.24(a) is shown in Figure 3.24(b).

EXAMPLE E25

For the magnetic circuit shown in Figure E25(a) calculate the flux density in the air gap when a current of 10 A flows through the coil. The material is cast steel with the B–H curve shown in Figure E25(b).

Solution

This is a case where the non-linear properties of the material cannot be ignored. Applying Ampere's law in the circuit of Figure E25(a) gives

$$NI = H_i l_i + H_g l_g.$$

Flux conservation gives the expression $B_i = B_g$ and hence $H_i = H_g/\mu_r$. However, since the material is non-linear μ_r varies depending on the value of the flux density. To proceed further the following procedure is adopted. A value of H_i is selected and the corresponding value of B_i is found from Figure E25(b). B_g is then found from flux conservation and $H_g = B_g/\mu_0$. These values are substituted in the expression for Ampere's law and different values are tried unitl the law is satisfied. A simple graphical construction where both sides of the equation are plotted on the same graph as a function of H_i is recommended. A reasonable starting value for H_i can be obtained by ignoring in the first instance the reluctance of the iron. Under these circumstances, $H_g \simeq NI/lg = 10^6\ \mathrm{A\,m^{-1}}$, $B_g = \mu_0 H_g = 1.256\ \mathrm{T}$, $B_i = B_g = 1.256\ \mathrm{T}$ and from Figure E25(b) $H_i = 1.4 \times 10^3\ \mathrm{A\,m^{-1}}$. Starting from this value, and following the procedure outlined above, results in a value for the flux density of $\sim 1.18\ \mathrm{T}$.

EXAMPLE E26

The B–H curve of a block of permanent magnet (pm) material is shown in Figure E26(a). Its cross-sectional area is $9\ \mathrm{cm}^2$ and its length is 5 cm. It is incorporated in the iron circuit of Figure E26(b) where the relative permeability of Fe is $\mu_r = 500$, the mean length of the magnetic path in Fe is 25 cm, the cross-sectional area of Fe is $9\ \mathrm{cm}^2$, the reluctance of the pole pieces is negligible, and 10% of the flux produced by the pm is leakage flux. Find the cross-sectional area and the

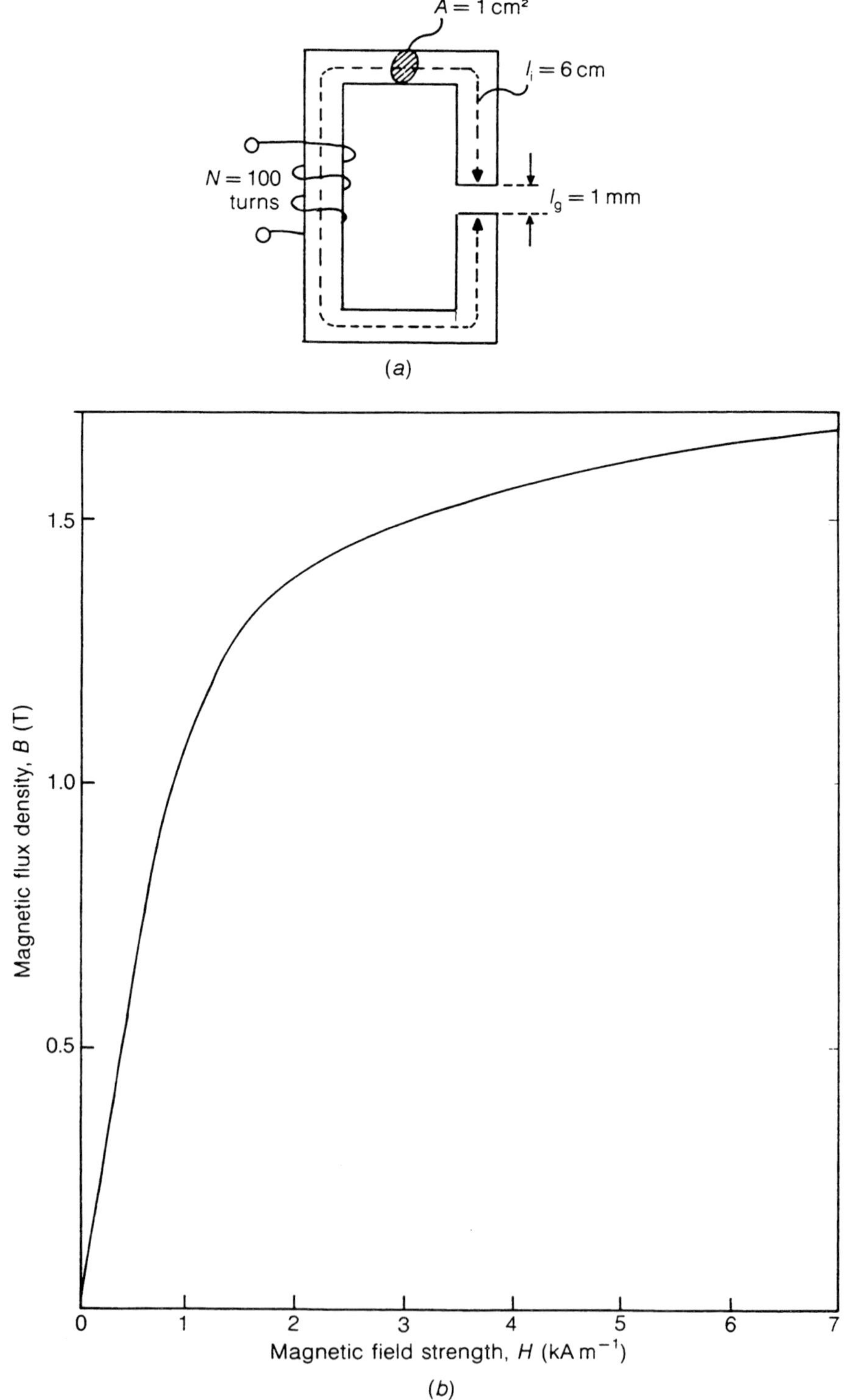

Figure E25 (*a*) Magnetic circuit with air gap. (*b*) Typical *B–H* curve for cast steel.

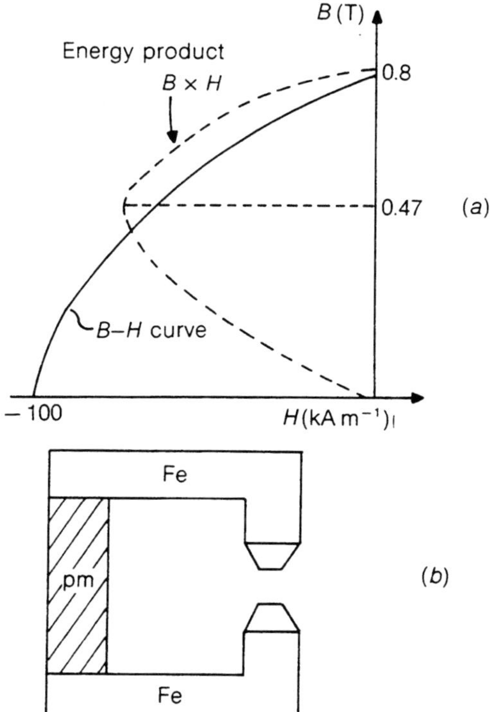

Figure E26

length of the air gap so that $B_g = 0.75\,\mathrm{T}$. *Hint*: operate the pm at its optimum point.

Solution

The energy product curve is plotted as the broken line in Figure E26(*a*) and the optimum operating point for the pm is found to be approximately $B_m = 0.47\,\mathrm{T}$, $H_m = -66 \times 10^3\,\mathrm{A\,m^{-1}}$. The flux through the pm is $0.47 \times 9 \times 10^{-4} = 4.23 \times 10^{-4}\,\mathrm{Wb}$. Due to leakage, the flux through the iron is $\phi_i = 0.9 \times 4.23 \times 10^{-4} = 3.8 \times 10^{-4}\,\mathrm{Wb}$. Since a flux density of 0.75 T is required in the gap, its cross-sectional area must be $A_g = 3.8 \times 10^{-4}/0.75 = 5.08\,\mathrm{cm^2}$. The flux density in the iron is $B_i = 3.8 \times 10^{-4}/(9 \times 10^{-4}) = 0.423\,\mathrm{T}$, hence $H_i = 0.423/(500 \times 4\pi \times 10^{-7}) = 673\,\mathrm{A\,m^{-1}}$. Similarly, $H_g = 0.75/(4\pi \times 10^{-7}) = 5.97 \times 10^5\,\mathrm{A\,m^{-1}}$. From Ampere's law, $-66 \times 10^3 \times 5 + 673 \times 25 + 5.97 \times 10^5 \times l_g = 0$ and hence $l_g = 0.525\,\mathrm{cm}$.

REMARKS

The full treatment of non-linear magnetic materials is very complex. In real networks, numerical techniques are almost always necessary.

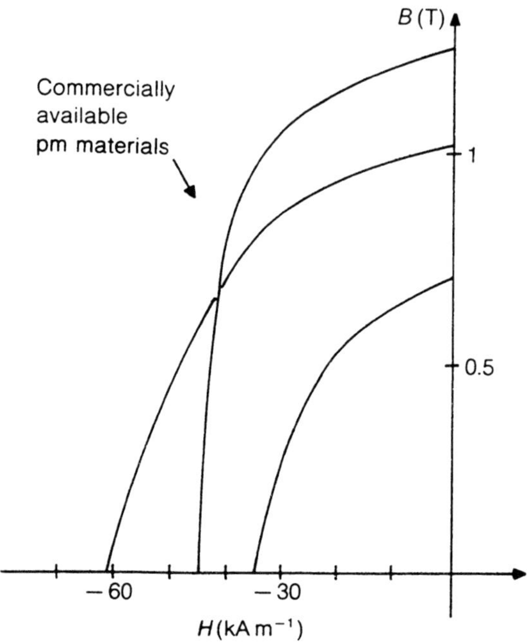

Figure 3.25 Typical B–H curves for commercially available permanent magnet materials.

Materials with a high energy product, suitable for making permanent magnets, are usually referred to as 'hard' magnetic materials to distinguish them from the 'soft' magnetic materials normally used in constructing magnetic circuits.

The B–H curve of some common pm materials is shown in Figure 3.25.

PROBLEMS

P37 The elements of a loudspeaker design are shown in Figure P37. The voice coil is mounted on a cone as shown. Item 1 is made out of mild steel ($\mu_r = 500$) whilst item 2 is made out of permanent magnet material. The pm curve is straight line defined by the points $B_r = 1$ T and $H_c = -5 \times 10^4$ A m^{-1}. After making the necessary simplifying assumptions, calculate the magnetic flux density produced by the pm in the air gap. If the voice coil is made out of wire of 1 m length, and is carrying a current of 125 mA, calculate the axial force acting on the voice coil due to the field produced by the pm.
(*Answer*: $B_m = 0.53$ T, force = 66 mN.)

P38 A magnetic circuit consisting of a permanent magnet (pm), soft iron, and air gap is shown in Figure P38(a). A typical B-line ABCDEFA is also shown. Leakage and fringing are neglected. The relevant part of the B–H curve is shown in

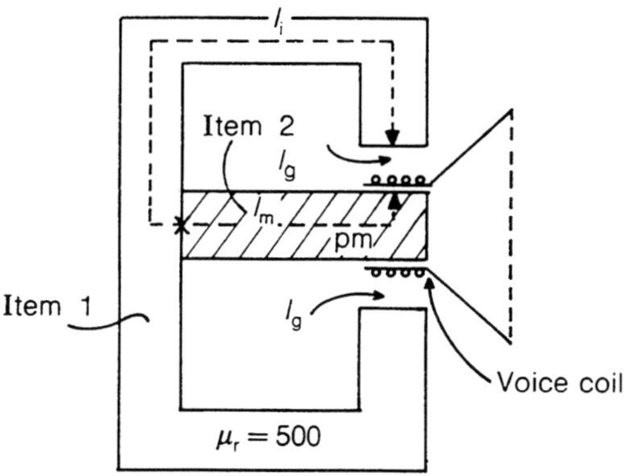

Cross-sections $A_i = A_g = A_m/2$
$l_i = 6$ cm, $l_m = 2$ cm, $l_g = 1$ mm

Figure P37

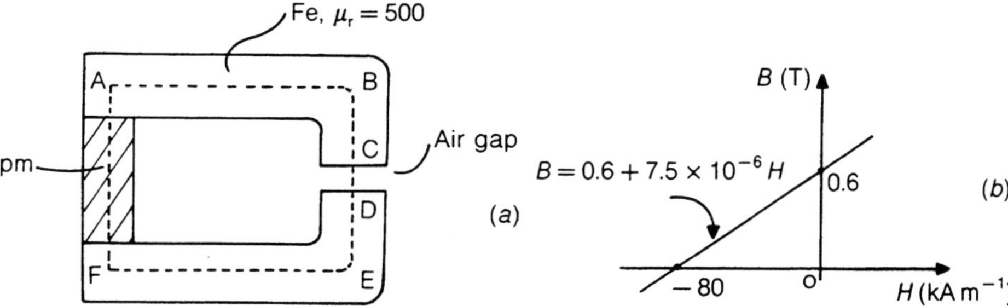

Figure P38

Figure P38(*b*). The dimensional details are:

	Length (m)	Cross-section (m^2)
AB	0.06	2×10^{-4}
BC	0.02	1×10^{-4}
CD	0.001	1×10^{-4}
DE	0.02	1×10^{-4}
EF	0.06	2×10^{-4}
FA	0.03	2×10^{-4}

Calculate the magnetic flux density in the gap.
(*Answer*: 0.8 T.)

P39 In the circuit shown in Figure E25(*a*) calculate the coil current required to establish a flux density of 0.82 T. Calculate the factor by which the current must increase to double the flux density to 1.64 T.
(*Answer*: 6.91 A, factor of 2.41.)

3.10 CONDITIONS FOR ELECTROMECHANICAL ENERGY CONVERSION

In section 3.8 expressions were derived which relate the force or torque acting on an electromechanical system to changes in magnetic energy stored. In particular, equation (3.20) was derived for a general type of machine consisting of fixed (stator) and moving (rotor) parts. This is repeated below:

$$T = \tfrac{1}{2} i_1^2 \frac{\mathrm{d}L_1}{\mathrm{d}\vartheta} + \tfrac{1}{2} i_2^2 \frac{\mathrm{d}L_2}{\mathrm{d}\vartheta} + i_1 i_2 \frac{\mathrm{d}M}{\mathrm{d}\vartheta}. \tag{3.20}$$

L_1 and L_2 are the self-inductance of the stator and rotor coils respectively and M is their mutual inductance. Several examples of systems were described, where a torque of magnetic origin develops. In this section a more systematic examination of electromechanical systems (motors and generators) is presented, based on equation (3.20).

An examination of (3.20) leads to the following conclusions:

(i) For torque to develop it is necessary that one or more of L_1, L_2 and M depend on the angle between rotor and stator.

(ii) It is possible for torque to be developed, even in the case when only one coil (on the rotor or stator) is energized, as long as the inductance of this coil depends on ϑ (first two terms in (3.20)). This is turn means that either the rotor or the stator should not be perfectly cylindrical. Non-cylindrical stators or rotors are referred to as 'salient', and the torque thus developed as 'reluctance torque'.

(iii) For the third term in (3.20) to contribute torque, both coils must be energized and their mutual inductance must depend on the angle between them. This contribution is sometimes referred to as 'excitation torque' and it is possible even in cases where both stator and rotor are perfectly cylindrical. In this case both L_1 and L_2 are independent of ϑ and the machine is referred to as a 'smooth air-gap machine'. Some examples of torque calculations were given in section 3.8. However, in constructing a useful electrical machine, development of torque over perhaps a limited part of the rotor's travel may not be sufficient for continuous rotation. An example is the case where torque develops over the whole range of values of ϑ, but in alternating directions so that, overall, the rotor is unlikely to move. Therefore, for a useful machine, more severe constraints must be imposed in addition to those evident from inspection of (3.20). To ensure continuous rotation it is necessary that the torque, when averaged over all angles ϑ, must not be equal to zero.

EXAMPLE E27

The stator of the machine shown in Figure E27 has a coil carrying a current $i_1 = I_1 \cos(\omega t)$. The salient rotor is not energized ($i_2 = 0$). Determine the conditions under which a non-zero average torque may be produced. Assume that $L = A + B\cos(2\vartheta)$ where A and B are constants.

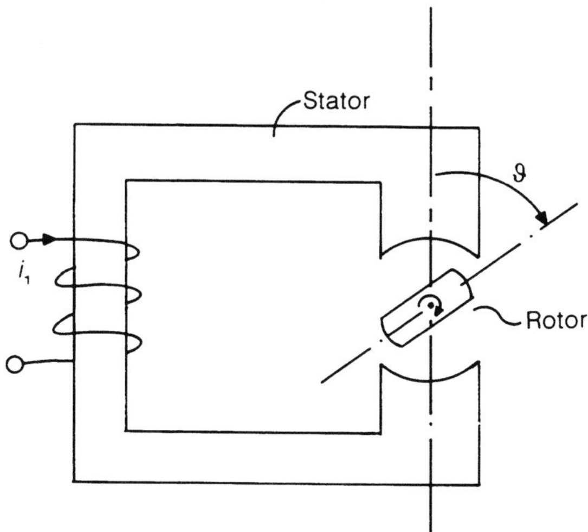

Figure E27

Solution

The expression for L is a reasonable one since, as expected, the inductance is maximum for $\vartheta = 0$ and minimum for $\vartheta = \pi/2$. Since i_2 is zero only the first term in (3.20) can possibly contribute to the torque. Substituting:

$$T_m = \tfrac{1}{2} I_1^2 \cos^2(\omega t)[-2B \sin(2\vartheta)] = -BI_1^2 \cos^2(\omega t) \cos(2\vartheta).$$

Suppose that the rotor rotates, as desired, at a constant speed ω_m. Then $\vartheta = \omega_m t + \delta$, where δ is the angle of misalignment between rotor and stator at the moment the rotor current is at positive maximum. Using trigonometric formulae:

$$\begin{aligned}\cos^2(\omega t)\sin[2(\omega_m t + \delta)] &= \tfrac{1}{2}(1 + \cos(2\omega t))\sin 2(\omega_m t + \delta)\\ &= \tfrac{1}{2}(\sin(2\vartheta)) + \tfrac{1}{4}\sin(2\vartheta + 2\omega) + \tfrac{1}{4}(\sin(2\vartheta - 2\omega t)).\end{aligned}$$

The time-averaged torque is defined as $\bar{T}_m = (1/T)\int_0^T T_m \mathrm{d}t$. Hence,

$$\bar{T}_m = -\tfrac{1}{2}BI_1^2\{\overline{\sin[2(\omega_m t + \delta)]} + \tfrac{1}{2}\overline{\sin[2(\omega + \omega_m)t + 2\delta]} + \tfrac{1}{2}\overline{\sin[2(\omega_m - \omega)t + 2\delta]}\}$$

where the overbar indicates a time average. The first term is clearly always zero, but the last two terms can contribute non-zero average torque when $\omega_m = \pm\omega$. In this case $\bar{T}_m = \pm\tfrac{1}{4}BI_1^2 \sin(2\delta)$. This is an example of a reluctance motor. For this type of machine electromechanical conversion can take place only when the speed of rotation is equal to the frequency of the currents in the stator (i.e. when there is synchronism). The simple motor shown is not self-starting.

EXAMPLE E28

Consider the device shown in Figure E28 (smooth air-gap machine), where the stator and rotor coils carry currents $i_1 = I_1 \cos(\omega_1 t)$ and $i_2 = I_2 \cos(\omega_2 t)$ respectively. Determine the conditions for average energy conversion.

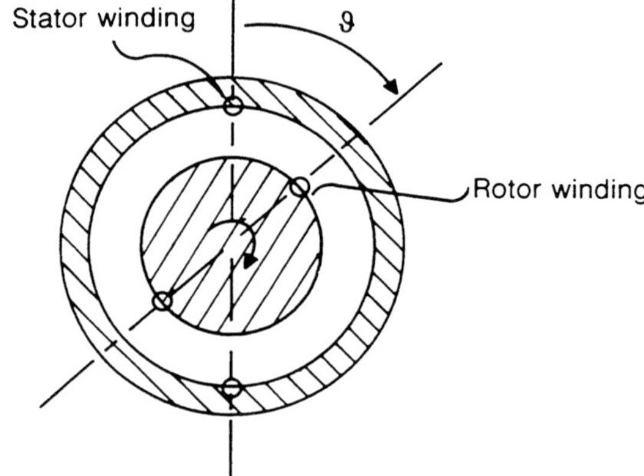

Figure E28

Solution

As before, the angle between the axes of the rotor and stator coils is equal to $\omega_m t + \delta$, where ω_m is the rotor angular speed and δ is the value of ϑ when the coil current is maximum. Clearly, because of cylindrical symmetry, L_1 and L_2 are independent of ϑ since the flux established by each coil is independent of the position of the other. However, mutual coupling is dependent on ϑ, and it is reasonable to assume that $M = M_0 \cos \vartheta$, where M_0 is a constant. Choosing $\delta = \pi/2$ then gives $M = M_0 \sin(\omega_m t)$. Only the third term in (3.20) can possibly contribute to the torque. Hence

$$\begin{aligned} T_m &= M_0 I_1 I_2 \cos(\omega_1 t) \cos(\omega_2 t) \cos(\omega_m t) \\ &= M_0 I_1 I_2 \cos(\omega_1 t) \tfrac{1}{2}\{\cos[(\omega_2 + \omega_m)t] \\ &\quad + \cos[(\omega_2 - \omega_m)t]\}. \end{aligned}$$

Clearly, a non-zero time average can only be achieved if a term without a sinusoidal time dependence emerges. Let us impose the condition $\omega_2 + \omega_m = \omega_1$; then

$$T = \tfrac{1}{2} M_0 I_1 I_2 \cos(\omega_1 t)\{\cos(\omega_1 t) + \cos[(\omega_2 - \omega_m)t]\}.$$

and the time-averaged torque is

$$\bar{T} = \tfrac{1}{2} M_0 I_1 I_2 \{\overline{\cos^2(\omega_1 t)} + \overline{\cos(\omega_1 t)\cos[(\omega_2 - \omega_m)t]}\}.$$

The second term in the brackets consists of cosine terms of $(\omega_1 + \omega_2 - \omega_m)t$ and $(\omega_1 - \omega_2 + \omega_m)t$ and hence it averages to zero. Only the first term contributes:

$$\bar{T} = \tfrac{1}{2} M_0 I_1 I_2 \tfrac{1}{2}\overline{[1 + \cos(2\omega_1 t)]} = \tfrac{1}{4} I_1 I_2 M_0.$$

Clearly, non-zero torque develops when the frequency of currents in the rotor (ω_2) plus the mechanical angular frequency of rotation (ω_m) equal the frequency of currents in the stator. This an example of a very popular type of machine, known as an induction motor. In a practical implementation of the machine, rotor currents are established by induction (there is no separate power supply to the rotor). On starting ($\omega_m = 0$) currents are induced in the rotor so that $\omega_2 = \omega_1$ and torque develops. As speed builds up, ω_m increases and hence ω_2 decreases but always remains equal to $\omega_1 - \omega_m$. At no load, $\omega_2 \simeq 0$ and $\omega_m \simeq \omega_1$. At full load, the mechanical speed ω_m drops below ω_1 so that just enough current is induced in the rotor to sustain the load.

REMARKS

The examples described above give a broad outline of the operating characteristics of basic machine types. Other examples are described in the problems. Practical machines based on these principles, have evolved over many years and incorporate many refinements to improve perfomance and efficiency. A description of these important features is beyond the scope of this introductory text.

PROBLEMS

P40 A smooth air-gap machine has a stator coil excited by a current $i_2 = I_2 \cos(\omega t)$ and a rotor coil excited by dc, $i_1 = I_1$. Assume that the mutual inductance between the coils is $M = M_0 \cos(\omega_m t + \delta)$, where ω_m is the mechanical speed of rotation and δ is a constant. Study the conditions for average electromechanical energy conversion, and calculate the average torque. Would a salient rotor increase the average torque? (Note: this is an example of a synchronous machine used for electricity generation in power stations.)
(*Answer*: $\omega_m = \omega$, $\bar{T} = \frac{1}{2} M_0 I_1 I_2 \sin\delta$. Yes–reluctance torque will develop.)

P41 Part of a device known as a dc machine is shown schematically in Figure P41. The stator and rotor coils are supplied by dc currents $i_1 = I_1$ and $i_2 = I_2$ respectively. The magnetic field pattern generated by the rotor appears stationary in spite of the rotation of the rotor. This is achieved by a device known as a commutator and the

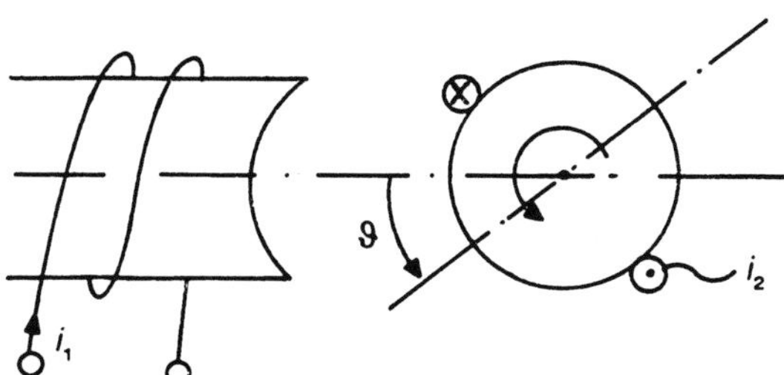

Figure P41

result is that the rotor can be represented by a stationary coil as shown. The mutual inductance between the two coils is $M = M_0 \cos \vartheta$. Obtain an expression for the torque and explain the operation of the machine when $\vartheta = \pm \pi/2$.
(*Answer*: $\bar{T} = -I_1 I_2 M_0 \sin \vartheta$, $\vartheta = +\pi/2$ for the generator, $\vartheta = -\pi/2$ for the motor).

P42 In the machine shown in Figure P42, the rotor inductance is given by the formula $L = L_0/[1 - 0.5 \cos (4\vartheta)]$ where L_0 is a constant and ϑ is the angle shown. The voltage source delivers a voltage equal to $v(t) = \omega A \sin (\omega t + \pi)$, where A is a constant and ω is the angular frequency.

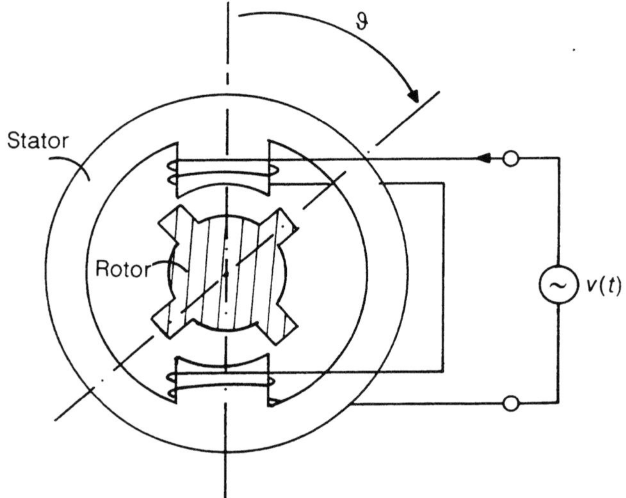

Figure P42

(*a*) Obtain a formula for the flux $\lambda(t)$ linked with the coil.
(*b*) For $\vartheta = \omega_m t + \delta$, obtain the conditions for average electromechanical energy conversion and derive a formula for the average torque.
(*Answer*: (*a*) $\lambda(t) = A \cos \omega t$, (*b*) $\omega_m = \pm \omega/2$, $T = -(A^2/4L_0) \sin (4\delta)$.)

PART II

PHENOMENA ASSOCIATED WITH SLOWLY ACCELERATING CHARGES

In Part I static electric and magnetic fields were described. Any changes in electrical parameters were assumed to be so slow that, essentially, electric and magnetic fields could be studied independently of each other. This assumption is not always valid. In fact, we were forced to accept this when the energization of a coil was considered. Faraday's Law, introduced in section 3.6, is an expression of the interdependence between electric and magnetic fields. In this second part we will examine in more detail this interdependence. We will also take the opportunity to study the relationship between field and circuit concepts. This exercise is useful in consolidating the understanding of fields by relating them to more familiar circuit concepts. Equally, our understanding of how practical circuit components behave at different frequencies will be considerably enhanced by reference to field concepts. This treatment will not be free from assumptions. It will be assumed that changes in electrical parameters are slow enough, and that the physical size of the circuit or other system studied in small enough, so that any interactions between different parts are practically instantaneous (quasistatic approxiation). Situations where the finite time required for the interaction between two parts of a system is important will be treated in Part III. If the period of variation of an electrical parameter is T, and the largest physical dimension of the system is s, then the time interval required for an interaction to be established is s/v_0, where v_0 is the propagation velocity for this interaction (the speed of light if this interaction is through air). The quasistatic approximation is valid provided that $T \gg s/v_0$.

Finally, the various mathematical models will be further developed and brought together into a small number of equations, known as Maxwell's equations, which describe in a compact and elegant form the whole range of electromagnetic interactions. Methods of solving these equations will also be described briefly.

CHAPTER 4

Interdependence Between Electric and Magnetic Fields

In cases where the acceleration of charges is significant, it is found that electric and magnetic fields are not independent. The generation of electric fields due to changing magnetic fields has already been described in section 3.6. If a magnetic flux is linked with a circuit it induces potential differences (and hence electric fields) given by Faraday's law:

$$V_{ind} = d\lambda/dt. \tag{3.14}$$

An example of the application of (3.14) is found in a device known as a transformer. In the magnetic circuit of Figure 4.1, two coils are wound as shown, and it is assumed that losses are negligible. A voltage source $v_1(t)$ is connected to coil 1 (referred to as the primary), whilst coil 2 (the secondary) remains an open circuit. The voltage $v_2(t)$ induced across the open-circuited terminals is required.

The energization of coil 1 results in a current i_1 and a flux $\phi(t)$ which will be calculated shortly. The flux linked with coil 1 due to current i_1 is then $\lambda_{11} = N_1\phi(t)$ and from Faraday's law the voltage induced across the terminals of coil 1 is $d\lambda_{11}/dt$. Since losses are neglected this voltage must be exactly equal to the applied voltage, i.e. $v_1(t) = d\lambda_{11}/dt = N_1 d\phi(t)/dt$. Similarly, the flux linked with coil 2 due to i_1 is $\lambda_{21} = N_2\phi(t)$ and hence the induced voltage on the secondary is $v_2(t) = d\lambda_{21}/dt = N_2 d\phi(t)/dt$. It follows therefore that $v_1/v_2 = N_1/N_2$; this is the well known transformer relation.

Let us now consider the general case when a load is connected across the secondary, so that a current i_2 can flow. To find the new induced voltages, Ampere's law is applied on curve C:

$$Hl = N_1 i_1 + N_2 i_2.$$

If $i_2 = 0$ (the special case just examined), $H = N_1 i_1/l$, and hence

$$\lambda_{11} = N_1\phi = N_1\mu\frac{N_1 i_1}{l}A = \mu\frac{N_1^2 A}{l}i_1 = L_1 i_1$$

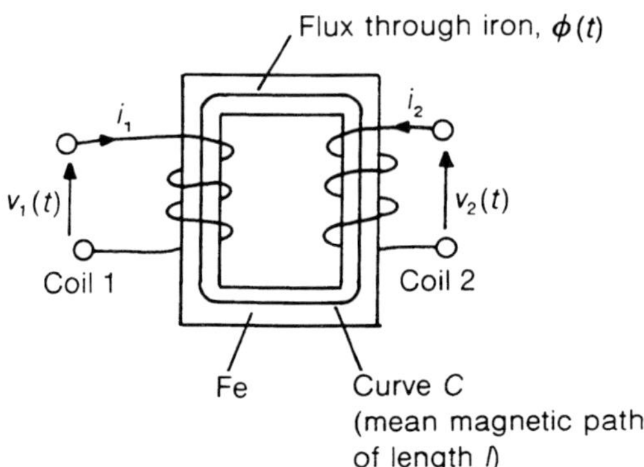

Figure 4.1 Basic transformer configuration.

where L_1 is the self-inductance of coil 1. In the general case ($i_2 \neq 0$), $H = N_1 i_1/l + N_2 i_2/l$ and $\phi = \mu N_1 A i_1/l + \mu N_2 A i_2/l$. The flux linked with coil 1 (due to both i_1 and i_2) is then

$$\lambda_1 = N_1\phi = \left(\mu\frac{N_1^2 A}{l}\right)i_1 + \left(\frac{\mu N_1 N_2 A}{l}\right)i_2.$$

The first quantity in brackets is the self-inductance L_1 of coil 1, and the second is the mutual inductance M between coils 1 and 2. Hence

$$\lambda_1 = L_1 i_1 + M i_2. \tag{4.1}$$

The first term describes the part of the flux due to current i_1 and the second term the flux due to i_2. Similarly the flux linked with coil 2 is found to be $\lambda_2 = N_2\phi = Mi_1 + L_2 i_2$. The voltages induced on each coil are:

$$v_1(t) = L_1\frac{di_1}{dt} + M\frac{di_2}{dt}$$

$$v_2(t) = M\frac{di_1}{dt} + L_2\frac{di_2}{dt}.$$

The reader may well enquire under what circumstances the familiar transformer relation $i_1/i_2 = N_2/N_1$ applies. From Ampere's law, $N_1 i_1 + N_2 i_2 = Hl$, and if it is assumed the μ_r is very high, then $H \simeq 0$. Under these circumstances $N_1 i_1 = -N_2 i_2$ (ampere × turn balance). The negative sign indicates that the two currents produce magnetic fields in opposition.

If a voltage $v_1(t) = V_{pk}\sin(\omega t)$ is applied, with the secondary open circuit, the flux builds up to a value $\phi(t) = \phi_{pk}\cos(\omega t)$ such that

$$V_{pk}\sin(\omega t) = N_1\frac{d}{dt}\phi_{pk}\cos(\omega t) = N_1\omega\phi_{pk}\sin(\omega t).$$

Hence,

$$\phi_{pk} = V_{pk}/N_1\omega. \tag{4.2}$$

This is sometimes known as the transformer design equation. The peak flux density in the iron is then $B_{pk} = V_{pk}/N_1\omega A$. To avoid saturation B_{pk} must be limited in practice to a value below approximately 1.5 T. If the applied voltage increases then so will B_{pk}. If this brings the iron into saturation then a high current will flow in the tansformer and damage will occur. A remedy is to increase the number of turns (N_1), or the cross-section of the iron (A), so that B_{pk} is kept below saturation. Which option is adopted in practice depends on whether losses in the wire (Cu losses) or losses in the magnetic circuit (Fe losses) are important. It should also be noted that if the option of increasing the frequency is available, B_{pk} can be kept below saturation with relatively low values of N_1 and A (i.e. with a small size transformer). This solution is adopted in mobile applications (e.g. transformers in aircraft).

Let us now focus attention on the related problem of a changing electric field being responsible for the production of a magnetic field. In the R–C circuit shown in Figure 4.2, a current will flow on closing the switch S and the capacitor will be charged. Since charge cannot flow between the plates, and charge must be conserved, the charge $i\mathrm{d}t$ flowing over time $\mathrm{d}t$ into the left-hand plate must be exactly equal to the increase $\mathrm{d}q$ in the charge on this plate. Hence $i = \mathrm{d}q/\mathrm{d}t$. At all times $V_s = iR + V_c$ (Kirchoff's voltage law). In steady state, $i = \mathrm{d}q/\mathrm{d}t = 0$ and the capacitor charges to a voltage V_s. The charge on the left-hand plate is then $Q = CV_s$ and on the right hand plate is $-Q$ (charge conservation). It is observed that during charging ($i \neq 0$) a magnetic field is established in the space between the plates. Applying Ampere's law on a contour such as C and assuming symmetry (approximately), leads to the conclusion that $\int_c H\mathrm{d}l = 0$ (since no free currents are enclosed) and hence $H = 0$. This is clearly a case where the mathematical model described so far is inadequate. The situation can be improved if it is postulated that between the capacitor plates a fictitious 'displacement' current flows which is equal to $A(\mathrm{d}\mathbf{D}/\mathrm{d}t)$. The term 'displacement' has been chosen for historical reasons. It will be shown that the displacement current (which is effectively the change in electric flux) accounts correctly for the magnetic field. It suffices to show that $A(\mathrm{d}D/\mathrm{d}t)$ is equal at all times to the current i and to modify Ampere's law to include, in addition to current due to free charges, the displacement current:

$$A\frac{\mathrm{d}D}{\mathrm{d}t} = A\varepsilon_0\frac{\mathrm{d}E}{\mathrm{d}t} = \frac{A\varepsilon_0}{d}\frac{\mathrm{d}V_c}{\mathrm{d}t} = \frac{A\varepsilon_0}{\mathrm{d}C}\frac{\mathrm{d}q}{\mathrm{d}t} = \frac{\mathrm{d}q}{\mathrm{d}t} = i.$$

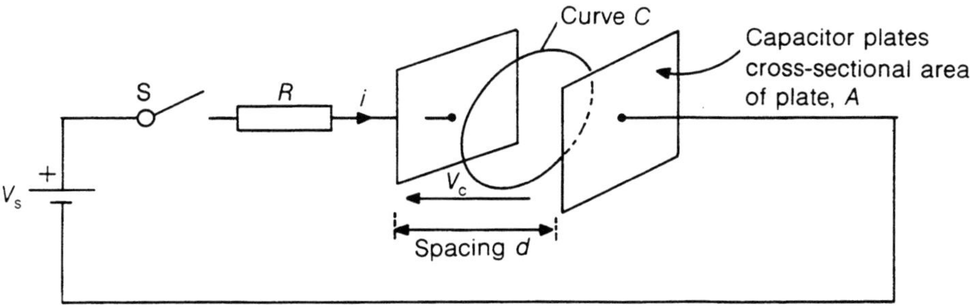

Figure 4.2 Capacitor charging and 'displacement current'.

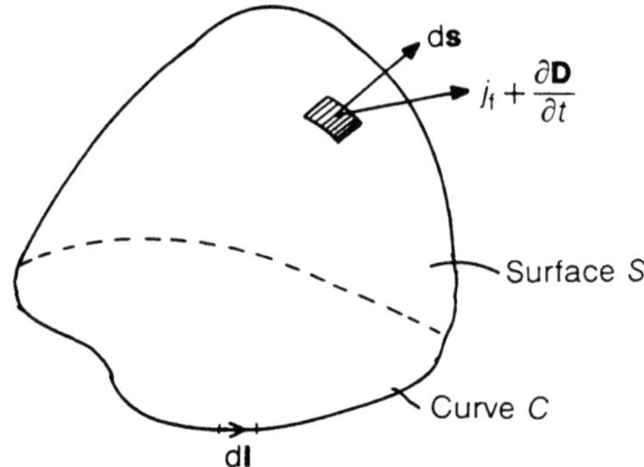

Figure 4.3 Configuration used for expressing Ampere's law.

The general form of Ampere's Law is now

$$\int_C \mathbf{H}\cdot d\mathbf{l} = \int_S \left(\mathbf{j}_f + \frac{\partial \mathbf{D}}{\partial t}\right) d\mathbf{s}. \tag{4.3}$$

The right-hand side of (4.3) is the total current (displacement + conduction) linked with curve C. In evaluating this current a surface S terminating on curve C is imagined as shown in Figure 4.3. The total current linked is evaluated by multiplying the current density $(\mathbf{j}_f + \partial \mathbf{D}/\partial t)$ by a small surface element $d\mathbf{s}$, and adding up all such contributions to cover the entire surface S (i.e. calculate the surface integral).

REMARKS

It is useful to generalize Faraday's law as follows. Let us consider the surface S terminating on curve C as shown in Figure 4.4. The flux linked with C is effectively the flux crossing surface S, i.e. $\int_S \mathbf{B}\cdot d\mathbf{s}$. The voltage induced around path C is $\int_C \mathbf{E}\cdot d\mathbf{l}$.

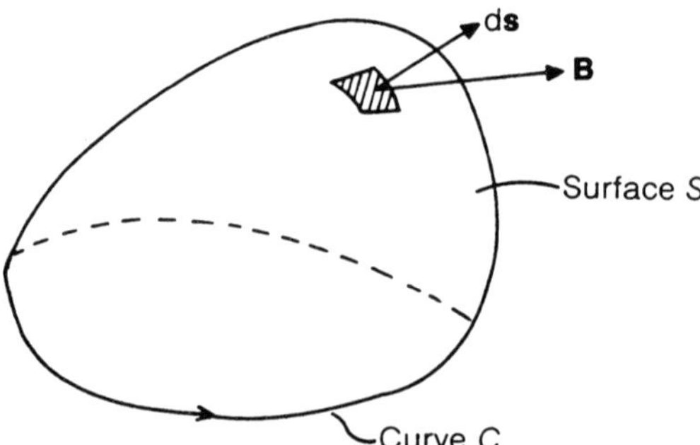

Figure 4.4 Configuration used for expressing Faraday's law.

Faraday's law can then be stated as:

$$-\frac{\partial}{\partial t}\int_S \mathbf{B}\cdot\mathbf{ds}=\int_C \mathbf{E}\cdot\mathbf{dl}. \tag{4.4}$$

If the curve C is traversed in a direction opposite to that shown in Figure 4.4, then the term on the left-hand side appears with a positive sign.

To summarize, magnetic fields are produced by currents due to free charges ($\mathbf{j}_f$ in (4.3)) and by changing electric fields $\partial\mathbf{D}/\partial t$ in (4.3)). Similarly, electric fields are produced by static charges, and also by changing magnetic fields (equation (4.4)). It should be noted that the contribution to the electric field of static charges does not appear in (4.4) since for a purely static field $\int_C \mathbf{E}\cdot\mathbf{dl}=0$.

PROBLEMS

P43 A transformer is shown in Figure P43. Losses and leakage are neglected. With the secondary open circuit, a voltage $v_1 = 100\cos(314t)$ is applied to the primary.

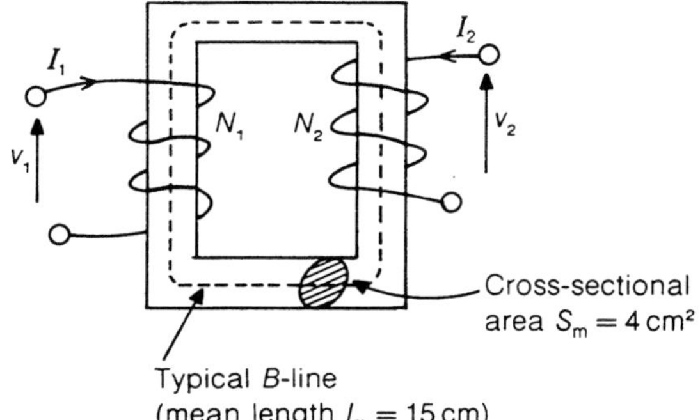

Figure P43

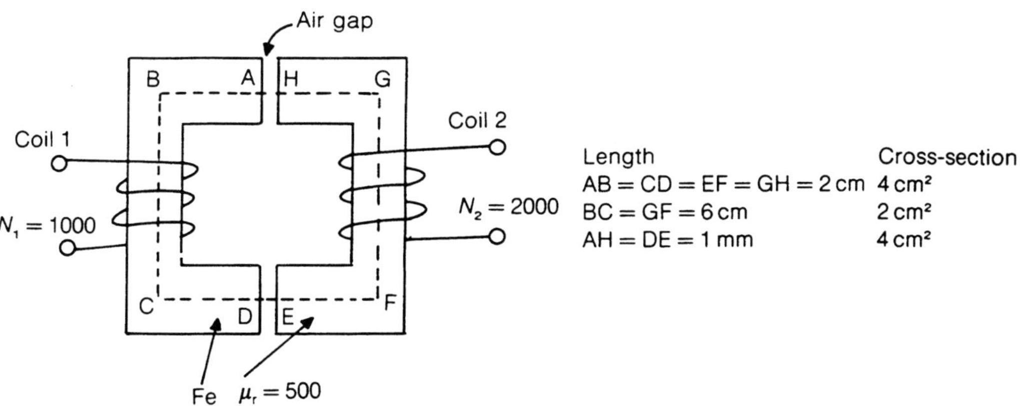

Figure P44

(*a*) Calculate the number of turns N_1 so that the peak flux density in the iron does not exceed 1 T.
(b) What will be the value of the magnetizing current I_1 in the two cases $\mu_r \to \infty$ and $\mu_r = 500$?
(c) Obtain the mutual inductance between the two coils when $N_2 = 20$ and $\mu_r = 500$.
(*Answer*: (a) 797 turns, (b) 0 A and 0.3 A, (c) 27 mH.)

P44 In the linear lossless magnetic circuit shown in Figure P44 leakage and fringing are neglected. Calculate:
(a) the inductance of coil 1;
(b) the inductance of coil 2;
(c) the mutual inductance between the two coils;
(d) the peak current which will flow though coil 1 when a voltage of rms value 50 V and frequency 50 Hz is applied to it whilst coil 2 is open circuit.
(*Answer*: (*a*) 0.19 H, (*b*) 0.76 H, (*c*) 0.38 H, (*d*) 1.17 A.)

CHAPTER 5

The Relationship Between Circuit and Electromagnetic Field Theories

The student and practitioner of electrical and electronic engineering can go a long way in understanding the behaviour of equipment and systems, and in designing such systems, by relying almost entirely on circuit concepts. It is therefore necessary to explain how circuit and field concepts are interelated and what benefits, if any, are brought about by introducing the concepts of the field which undeniably adds yet another level of complexity and abstraction.

5.1 LUMPED PARAMETER CIRCUITS

Equivalent circuits describing 'real' components and systems are constructed from a small number of 'ideal' components, as follows.

(i) Ideal sources (e.g. voltage/current sources), capable of supplying electrical energy at the expense of some other form of energy (e.g. chemical energy in a battery, potential energy of water in a hydroelectric generator).

(ii) An ideal energy-dissipation element known as a resistor, where electrical energy is converted into heat.

(iii) Two ideal energy-storage elements, namely, the capacitor and the inductor.

In an ideal capacitor, energy storage is associated with charge separation (a form of 'potential' energy) and hence with storage in the electric field system. In an ideal inductor, energy storage is associated with charge motion (a form of 'kinetic' energy) and hence storage in the magnetic field system. Even for ideal circuit components the value of capacitance or inductance is intimately related to the structure of the electromagnetic field associated with the component. Moreover, examination of the fields associated with real components reveals that it is almost never true that such components exhibit purely capacitive or purely inductive

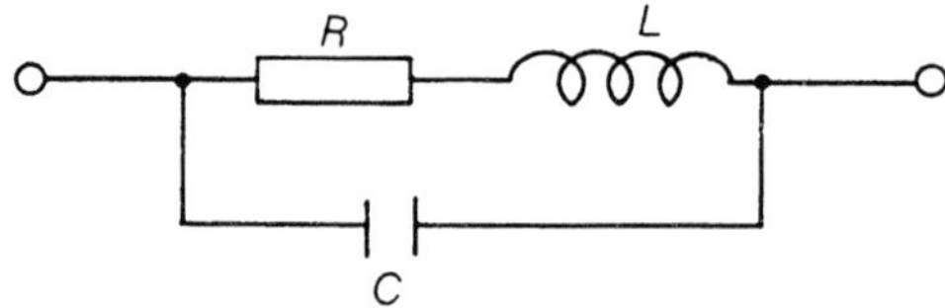

Figure 5.1 General circuit model of a real component.

behaviour. A piece of wire coiled to form an inductor will in use develop potential differences and hence an electric field in its proximity. This in turn implies energy storage in the electric field and hence some capacitive behaviour. Equally, a capacitor will in use experience current flow and hence a magnetic field will be created. It will display therefore some inductive behaviour. It will also be shown that losses are associated with the establishment of electric and magnetic fields. These are in addition to ohmic losses in wires. It is therefore reasonable to assume that any real component can be constructed by combining ideal components as shown in Figure 5.1. This arrangement is one of many which can be used to describe real components. Physical intuition may suggest others. The value of these components can only be determined by resorting to field considerations as shown in Chapters 2 and 3. In general they depend strongly on frequency, and sometimes also on the value of the applied voltage or current (non-linear components).

An example of how the nature of a circuit component changes with frequency can be seen in the behaviour of a typical commercial capacitor (0.25 μF with 10 cm leads). Such a component above approximately 1 MHz shows strong inductive behaviour. Cutting down the leads to a length of less than 1 cm can increase this frequency to 2 MHz. In contrast, a coaxial capacitor (feed-through type) retains its capacitive character up to a frequency of a few tens of MHz.

All these aspects of real component behaviour can be better understood by reference to field concepts. This will be done in the next few sections. It is, however, worth emphasizing that the concentration (lumping) of specific behaviour (e.g. losses, capacitance) on physically distinct components, as shown in Figure 5.1, is in itself an approximation. It is only justifiable, if the time delay required for an electrical disturbance to propagate across the largest physical dimension of the component or system is much smaller than the characteristic time of the electrical disturbance (e.g. its period). A more detailed examination of the circumstances under which lumped component representation of a system is permissible will be given in section 5.4. For the remainder of this section the interrelationship between lumped component circuit and field fundamental laws will be explored.

Let us consider the conducting loop shown in Figure 5.2, driven by a voltage source and placed in a region where an externally generated magnetic field is established. Let the magnetic field linked with the loop be designated by ϕ_i, where the subscript i stands for incident, i.e. coupled with the loop from external influences. Faraday's law (4.4) indicates that the total voltage induced around the loop is $\int_{\text{loop}} \mathbf{E} \cdot \mathrm{d}\mathbf{l} = \mathrm{d}\phi/\mathrm{d}t$, where ϕ is the *total* magnetic flux linked with the loop. Let us first evaluate the left-hand side of this equation. Integration across the source (path

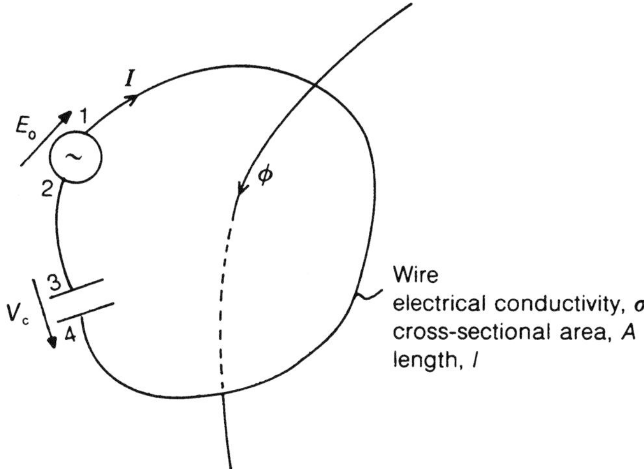

Figure 5.2 Generalized Kirchhoff's voltage law.

1 to 2) gives $\int_1^2 \mathbf{E}\cdot d\mathbf{l} = E_0$. Integration along paths 2 to 3 and 4 to 1 gives

$$\int \mathbf{E}\cdot d\mathbf{l} = \int \frac{-j}{\sigma} dl = \int \frac{-(I/A)}{\sigma} = -I\frac{l}{\sigma A} = -IR,$$

where R is the resistance of the wire. In this manipulation it has been assumed that the current is distributed uniformly over each wire section ($j = I/A$). If this is not the case corrections to the value of R must be made. Finally integration across the capacitor (path 3 to 4) gives $\int_3^4 \mathbf{E}\cdot d\mathbf{l} = -V_c$. Hence in total,

$$\int_{\text{loop}} \mathbf{E}\cdot d\mathbf{l} = E_0 - IR - V_c.$$

Let us now examine the right-hand side of Faraday's law. The total flux ϕ linked with the loop consists of two parts: Firstly, the externally produced flux ϕ_i and secondly the flux ϕ_s produced by any current flowing in the loop. The subscript s stands for 'scattered' to indicate the flux produced by the current in the loop. Hence ϕ is equal to $\phi_i + \phi_s$. The flux produced by the loop current is proportional to the current I, and the constant of proportionality is, by definition, the inductance L of the loop, i.e. $\phi_s = LI$. Therefore

$$\frac{d\phi}{dt} = \frac{d\phi_i}{dt} + \frac{d\phi_s}{dt} = \frac{d\phi_i}{dt} + L\frac{dI}{dt}.$$

Faraday's law for the loop then gives

$$E - IR - V_c = \frac{d\phi_i}{dt} + L\frac{dI}{dt} \qquad \text{or} \qquad E - \frac{d\phi_i}{dt} = IR + L\frac{dI}{dt} + V_c.$$

This expression is none other than Kirchhoff's voltage law, familiar from circuit theory. The expression on the left-hand side is the total driving voltage for the loop consisting of the voltage source and magnetic induction term.

PROBLEM

P45 A node formed by three branches is shown in Figure P45. The surface S_1 cuts throught R and C as shown. The inductor penetrates this surface through a hole. Apply Ampere's law (4.3) on the curve c defining the hole, and show that it reduces to Kirchoff's current law at this node. *Hint*: evaluate the right-hand side of (4.3) on surface S_1 and on surface S_2, and equate these two expressions. Take particular care with the signs of the various terms.

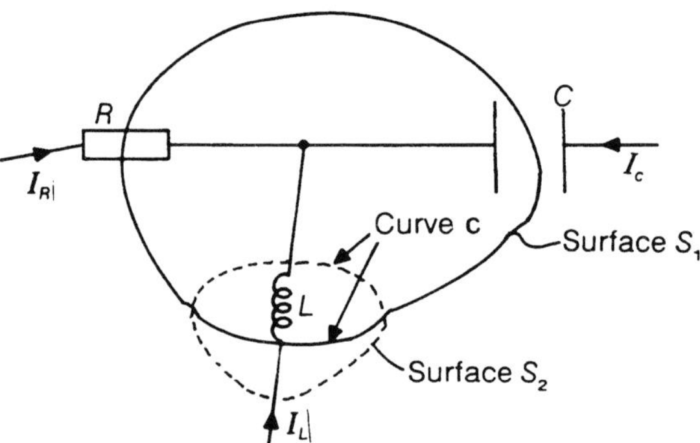

Figure P45

5.2 CURRENT DISTRIBUTION INSIDE CONDUCTORS

Of crucial importance in determining the behaviour of circuits and electromagnetic devices is the understanding of the manner in which current distributes inside conductors. The term conductors is used in the broader sense to include, apart from copper wires, also any other metallic body in which current may be induced (e.g. iron parts in machines or inductors).

The kind of problems addressed in this section are the following. If a coil, supplied by a high-frequency source, is wound on a solid iron former, is it reasonable to assume that the flux density distributes uniformly in the cross-section of the iron? If an ac current flows through a copper wire, is it correct to assume that the current is distributed uniformly in its cross-section? Answers to these questions will be given below together with quantitative means of assessing the significance of the results.

Let us consider the half-space $z > 0$ (Figure 5.3) filled with a material $\mu_r\mu_0$, whilst the half-space $z < 0$ consists of air. A magnetic field is established in the air along the y-direction so that $H_y(z = 0) = H_0 \cos(\omega t)$. Although the configuration in Figure 5.3 is impractical, it approximates fairly well the case of a coil wound on a large-diameter iron core. The field H_y is set up in the air by the coil, and the problem posed

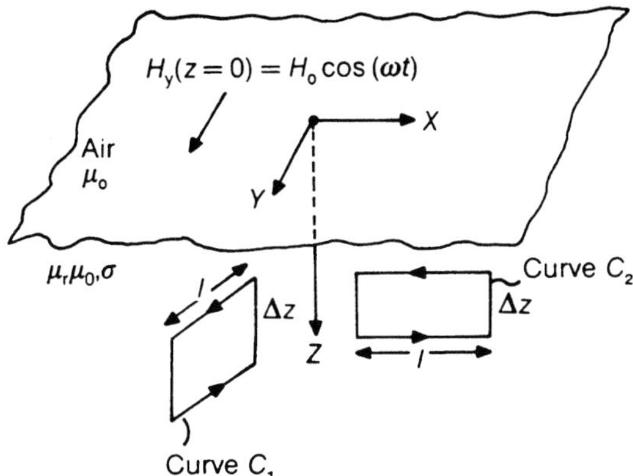

Figure 5.3 Magnetic field penetration into a semi-infinite block of magnetic material.

is to determine how this field penetrates into the iron. It is to be expected that the changing magnetic flux will induce voltages and hence currents inside the iron. From the symmetry of the problem, and since any such current opposes the applied field (Lenz's law), it follows that the current must be flowing in the x-direction. Similarly the induced electric field is also in the x-direction. These currents driven by the induced voltages are known as 'eddy currents' and affect significantly the design and operation of power and high-frequency devices. Let us now pursue further these ideas to obtain quantitative results.

Curve C_1 is chosen as shown in Figure 5.3 and Ampere's law is applied around it, $-(H_y + \Delta H_y)l + H_y l = j_x l \Delta z$, where j_x is the current density in $\mathrm{A\,m^{-2}}$ at the location of loop C_1. Simplifying this expression and taking limits as Δz gets small gives

$$\frac{\mathrm{d}H_y}{\mathrm{d}z} = -j_x \qquad \text{or} \qquad \frac{1}{\mu}\frac{\mathrm{d}B_y}{\mathrm{d}z} = j_x. \tag{5.1}$$

Faraday's law applied on C_2 gives

$$-E_x l + (E_x + \Delta E_x)l = -\frac{\mathrm{d}}{\mathrm{d}t}(B_y l \Delta z).$$

But $\Delta E_x = \Delta j_x/\sigma$, hence

$$\frac{1}{\sigma}\frac{\mathrm{d}j_x}{\mathrm{d}z} = -\frac{\mathrm{d}B_y}{\mathrm{d}t}. \tag{5.2}$$

Equations (5.1) and (5.2) can in principle be solved, subject to appropriate boundary conditions, to obtain the two unknown quantities j_x and B_y inside the material. This task is made easier if account is taken of the fact that, in a linear medium, all quantities will vary with the frequency of the source. *Phasors* can therefore be used to simplify the solution procedure. For every quantity $f(z,t)$ a complex function $\bar{f}(z)$ is defined in such a way that $f(z,t) = \mathrm{Re}\{\bar{f}(z)\mathrm{e}^{\mathrm{j}\omega t}\}$, where Re stands for the real part. The expression in the brackets is called the phasor representing $f(z,t)$. Usually only $\bar{f}(z)$ is used in calculations with the exponential term added at the end. For example, the complex quantity representing the magnetic field intensity at the surface is $H_0 + j0$. Indeed, $\mathrm{Re}\{(H_0 + j0)\, e^{j\omega t}\} = H_0 \cos(\omega t)$. It follows therefore that in phasor notation

$$\frac{1}{\mu}\frac{\mathrm{d}\bar{B}_y}{\mathrm{d}z} = -\bar{j}_x \tag{5.3}$$

$$\frac{1}{\sigma}\frac{\mathrm{d}\bar{j}_x}{\mathrm{d}z} = -j\omega\bar{B}_y \tag{5.4}$$

Differentiating (5.3) with respect to z and substituting gives

$$\frac{1}{\mu}\frac{\mathrm{d}^2\bar{B}_y}{\mathrm{d}z^2} = -\frac{\mathrm{d}j_x}{\mathrm{d}z} = \mathrm{j}\omega\sigma\bar{B}_y$$

or

$$\frac{\mathrm{d}^2\bar{B}_y}{\mathrm{d}z^2} = \mathrm{j}\omega\mu\sigma\bar{B}_y = k^2\bar{B}_y \tag{5.5}$$

where $k = \sqrt{\mathrm{j}\omega\mu\sigma} = \sqrt{\mathrm{j}}\sqrt{\omega\mu\sigma}$. But $\mathrm{j} = \mathrm{e}^{\mathrm{j}\pi/2}$ and $\sqrt{\mathrm{j}} = \mathrm{e}^{\mathrm{j}\pi/4} = (1+\mathrm{j})/\sqrt{2}$, hence

$$k = (1+\mathrm{j})\sqrt{\omega\mu\sigma/2} = (1+\mathrm{j})/\delta.$$

The quantity $\delta = \sqrt{2/\omega\mu\sigma}$ has a special physical significance as will be shown shortly. The solution of (5.5) is straightforward:

$$\bar{B}_y = A_1 e^{kz} + A_2 \mathrm{e}^{-kz}$$

where A_1, A_2 are constants to be determined from the boundary conditions. Substituting for k gives

$$\bar{B}_y = A_1 \mathrm{e}^{z/\delta}\mathrm{e}^{\mathrm{j}z/\delta} + A_2 \mathrm{e}^{-z/\delta}\mathrm{e}^{-\mathrm{j}z/\delta}.$$

The first term tends to infinity as z increases which is physically unacceptable; hence $A_1 = 0$, and

$$\bar{B}_y = A_2 \mathrm{e}^{-z/\delta}\mathrm{e}^{-\mathrm{j}z/\delta}.$$

Therefore $B_y(z,t) = \mathrm{Re}\{\bar{B}_y \mathrm{e}^{\mathrm{j}\omega t}\} = A_2 \mathrm{e}^{-z/\delta}\cos(\omega t - z/\delta)$. At $z = 0$, $B_y = B_0$ and hence $A_2 = B_0$. Substituting,

$$B_y(z,t) = B_0 \mathrm{e}^{-z/\delta}\cos(\omega t - z/\delta). \tag{5.6}$$

The total flux ϕ_y in the iron crossing a surface extending from $z = 0$ to infinity and for a metre length along the x-direction is

$$\phi_y = \int_{z=0}^{\infty} B_y(z)\,\mathrm{d}z = \mathrm{Re}\left\{\int_{z=0}^{\infty} B_0 \mathrm{e}^{\mathrm{j}(\omega t - z/\delta) - z/\delta}\mathrm{d}z\right\}$$

$$= \mathrm{Re}\left\{B_0 \mathrm{e}^{\mathrm{j}\omega t}\int_{z=0}^{\infty} \mathrm{e}^{-\mathrm{j}z/\delta - z/\delta}\mathrm{d}z\right\}$$

$$= \mathrm{Re}\left\{B_0 \mathrm{e}^{\mathrm{j}\omega t}\frac{\mathrm{e}^{-\mathrm{j}z/\delta - z/\delta}}{-\mathrm{j}/\delta - 1/\delta}\Bigg|_{z=0}^{\infty}\right\}.$$

Finally, after some algebra

$$\phi_y = \frac{B_0\delta}{\sqrt{2}}\cos(\omega t - \pi/4). \tag{5.7}$$

Hence the flux lags the magnetic field at the surface (and hence the coil current) by an angle $\pi/4$.

The current density inside the iron can also be obtained easily:

$$\bar{j}_x = -\frac{1}{\mu}\frac{\mathrm{d}\bar{B}_y}{\mathrm{d}z} = \frac{1}{\mu}B_0(1+j)\frac{1}{\delta}\mathrm{e}^{-jz/\delta - z/\delta}$$

and

$$j_x(z,t) = \frac{\sqrt{2}B_0}{\mu\delta}\mathrm{e}^{-z/\delta}\cos(\omega t - z/\delta + \pi/4). \tag{5.8}$$

Expressions (5.7) and (5.8) indicate that the peak value of magnetic flux and current density decrease away from the surface and into the magnetic material. The magnetic flux penetrates only into a thin surface layer. The characteristic depth of penetration (depth where the flux density drops to 1/e of its value at the surface) is called the skin depth:

$$\delta = \sqrt{\frac{2}{\mu\omega\sigma}}. \tag{5.9}$$

At high frequencies or when the electrical conductivity is high, the skin depth is very small and magnetic field penetration inside the material is very difficult. For example, in iron ($\sigma = 5 \times 10^8\,\Omega^{-1}\,\mathrm{m}^{-1}$, $\mu_r \simeq 500$) and at a frequency of 50 Hz. $\delta \simeq 1.4$ mm. Under these circumstances it is evident that the construction of iron parts in thicknesses larger than a few skin depths (say 3 or 4 mm) will offer no advantages in providing a low-reluctance path. Any non-uniform distribution of a quantity, such as magnetic flux density or current density, implies an inefficient use of the material. In the case of $\bar{B}$ it means a high-reluctance path (since only a few layers near the surface carry flux) and in the case of $\bar{j}$ increased ohmic losses (since the current flows over a restricted part of the cross-section of the material).

A coil wound on an iron former under circumstances where δ is much smaller than the diameter of the former will have a lower inductance and higher equivalent resistance compared with the values expected if the whole cross-section of the former was exploited. This situation is clearly undesirable and if not remedied would make the design of efficient machinery or high frequency devices impossible. To overcome these difficulties it is necessary to prevent, as far as possible, the flow of eddy currents. This can be achieved by laminating the material as shown in Figure 5.4, where thin sheets of the material, treated in such a way that a small amount of insulation is formed on their surface, are stacked together to build up any desired shape. They can be arranged with respect to the magnetic field as in either Figure 5.4 (*a*) or (*b*). Eddy currents must flow, as shown schematically, so that they oppose the magnetic field. Clearly the presence of insulation in Figure 5.4(*a*) restricts the flow of eddy currents, whilst in (*b*) it has a minimal effect. Introducing laminations as in (*a*) significantly reduces eddy currents and results

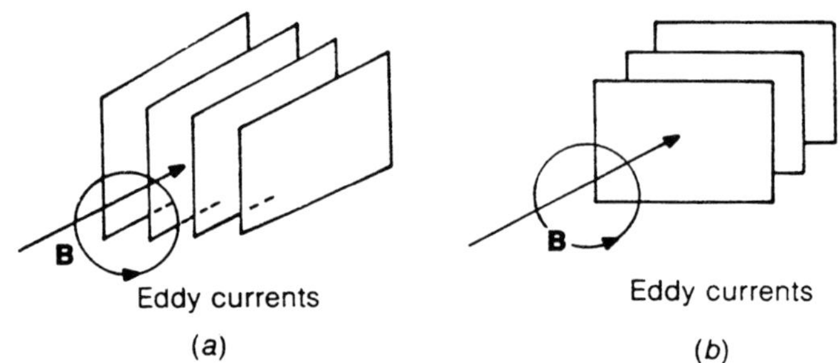

Figure 5.4 Effective (*a*) and ineffective (*b*) use of laminations

in a more uniform distribution of flux and lower losses. In devices where high frequency or rapidly changing fields are present, laminations are always used.

It remains to determine how thick the laminations should be. Clearly a thickness of the order of the skin depth δ is appropriate. As the frequency increases δ becomes smaller and it becomes impractical to construct very thin laminations. Instead, at high frequencies, magnetic materials (ferrites) are constructed out of small grains using a sintering process. Although discussion so far has been based on the penetration of flux inside iron, similar conclusions are drawn when the distribution of current inside a conductor is considered. When an ac current flows through a cylindrical conductor, a magnetic field is established in the conductor. Axial currents will be induced so that the flux inside the conductor is minimized. The magnetic field at the surface and outside the conductor is the same, irrespective of the actual distribution of the current (assuming that symmetry is maintained). Therefore, to minimize the field inside the conductor, the field, and hence most of the current, must be concentrated in a thin layer, of the order of a skin depth δ, near the surface of the conductor. For copper and at a frequency of 50 Hz, $\delta \cong 10\,\text{mm}$. Hence in conductors with cross-sectional dimensions exceeding this value the current distributes non-uniformly over the cross-section. At very high frequencies only the surface layer is active in current conduction. This has two important consequences. Firstly, the effective resistance of the conductor at high frequencies is higher when compared with the resistance at dc. At dc, a conductor of radius a has an effective cross-section πa^2 (since the current distributes uniformly over the cross-section). At high frequencies the effective cross-section is reduced to approximately $2\pi a\delta$ (provided that $\delta \ll a$) and hence the resistance is higher. As the operating frequency increases this results in higher losses and reduced efficiency. Secondly, the presence of eddy currents reduces the magnetic flux inside the conductor and so the inductance decreases as the frequency is increased. For example, the inductance per unit length of a cylindrical conductor assuming current flow near the surface (high-frequency case) is determined by the flux external to the conductor. If the same current is distributed uniformly over the cross-section (low-frequency case) then to the external inductance mentioned above must be added an extra term $\mu_0/4\pi$, the so-called internal inductance, (see Problem P27). Eddy currents in conductors can be minimized by building up large cross-sections using several strands of thin wire—a practice akin to the use of laminations.

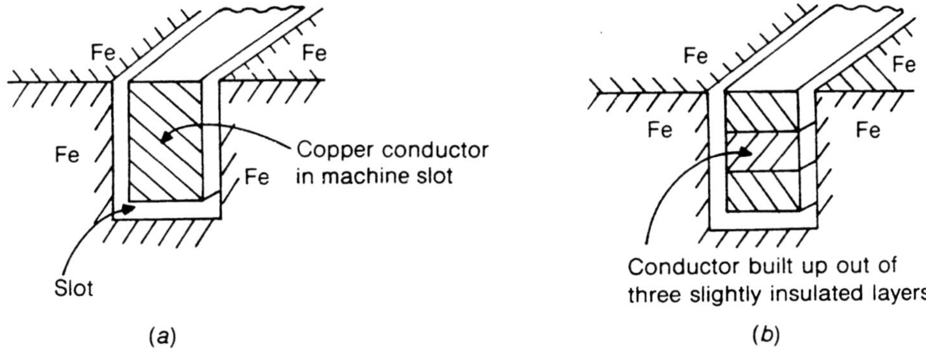

Figure 5.5 (a) Solid Cu conductor in an Fe slot and (b) segmented conductor.

Another example of the significance of eddy currents is shown in Figure 5.5(a), where a conductor of rectangular cross-section is placed inside the slot of an electrical machine. Eddy-current flow is such that the current is concentrated near its upper surface (near the air). This increases losses and affects the operational behaviour of the machine. The same conductor cross-section can be built up by three smaller conductors, slightly insulated from each other, as shown in Figure 5.5(b). These are, effectively, twisted together (a process known as transposition) so that each of the three conductors occupies on average the same position in the slot. In this way a more uniform current distribution can be achieved.

EXAMPLE E29

In the arrangement shown in Figure E29, a coil (N turns) is wound on an iron core. The dimensions of the iron and the insulating layer (air) between the coil and the core are indicated in the figure, as is the frequency of the current supplied to the coil. Obtain the parameters of the coil equivalent circuit as seen across terminals A, B.

Solution

This is an example where the understanding of field concepts will lead to a suitable circuit model which can then be used further in designing the coil and its power supply.

In tackling this problem several assumptions will be made:

(i) It is assumed that the coil appears effectively as a copper layer of the cross-section shown and that the current is concentrated on its inner surface. This is a reasonable assumption when $\delta_{Cu} \ll d_{Cu}$.

(ii) Similarly, the magnetic flux and eddy currents are constrained on the outer surface of the iron core.

(iii) End effects are neglected ($d_{Cu} \ll l$).

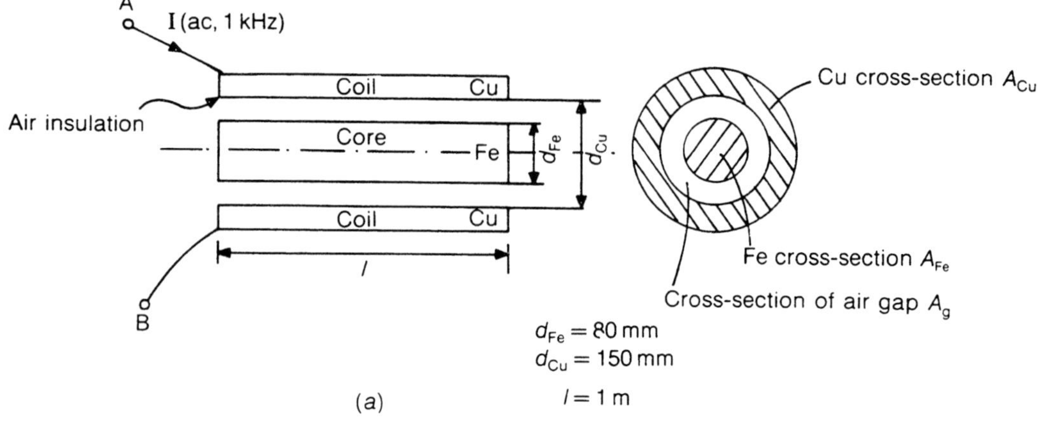

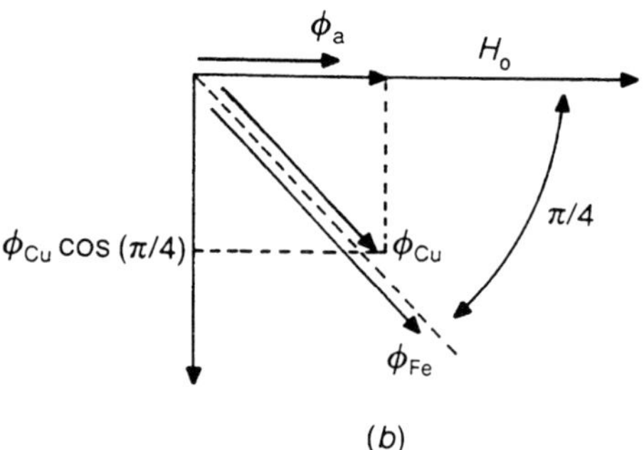

Figure E29 (a) Cylindrical coil wound around a cylindrical core and (b) phasor diagram

Following these assumptions, the physical picture is of flux ϕ_{Cu} on a thin layer of the coil, ϕ_a in the insulation layer, and ϕ_{Fe} in a thin layer of the iron. The total flux linked with the coil is the combination of ϕ_{Cu}, ϕ_{Fe} and ϕ_a. The magnitude of this flux and its phase relative to the coil current must now be calculated. Let the rms value of the magnetic field intensity in the insulation layer H_0 be chosen as the reference as shown in Figure E29(b). Then ϕ_{Cu} and ϕ_{Fe} lag $\bar{H}_0$ by $\pi/4$ as already shown, whilst ϕ_a is in phase with $\bar{H}_0$. From (5.7) the rms value is obtained $\phi_{Cu} = \mu_0 H_0 \pi d_{Cu} \delta_{Cu}/\sqrt{2}$ and its real and imaginary parts are $\mu_0 H_0 \pi d_{Cu} \delta_{Cu}/2$ as shown. Similar considerations apply for ϕ_{Fe}. Hence using phasor notation, the total rms flux is

$$\bar{\phi} = \mu_0 H_0 \left\{ \left[\mu_r \frac{\pi d_{Fe}}{2} \delta_{Fe} + \frac{\pi d_{Cu}}{2} \delta_{Cu} + A_g \right] + j \left[\mu_r \frac{\pi d_{Fe}}{2} \delta_{Fe} + \frac{\pi d_{Cu}}{2} \delta_{Cu} \right] \right\}$$

where $A_g = \pi(d_{Cu}^2 - d_{Fe}^2)/4$.

The value of H_0 is related to the coil current I through Ampere's law, namely $H_0 = NI/l$. Similarly, using phasors, the coil voltage is obtained from Faraday's law: $\bar{V} = j\omega\bar{\phi}$. Substituting in this formula gives

$$\bar{V} = \frac{\mu_0 I N^2 \omega}{l}[A + j(A + A_g)]$$

where

$$A = \mu_r \frac{\pi d_{Fe}}{2}\delta_{Fe} + \frac{\pi d_{Cu}}{2}\delta_{Cu}.$$

Across terminals A and B the coil plus the iron piece appear as an R–L series combination where

$$R = \frac{\mu_0 N^2 \omega A}{l} \qquad (\Omega)$$

$$L = \frac{\mu_0 N^2 (A + A_g)}{l} \qquad (\text{H}).$$

The reader should make numerical substitutions to get a feel for these parameters.

REMARKS

Current and magnetic field do not distribute uniformly inside conducting materials. This is a consequence of the flow of eddy currents. A more uniform distribution is obtained by inhibiting the flow of eddy currents (by lamination in iron or stranding in conductors). The high-frequency resistance is higher compared to the low-frequency resistance. This, in general, is undesirable in the design of devices, but it may be used with advantage when wishing to heat only the surface of conducting bodies (e.g. induction heating for the surface treatment of steel components). The high-frequency inductance is always smaller when compared with its value at low frequencies. Proper design can minimize variations with frequency but it cannot eliminate them. In cases where several conductors are near to each other, further distortions of current distribution are caused (the proximity effect).

In applying Ampere's law in this section, it was assumed that the current linked is the conduction current only, and that the contribution of the term $\partial\bar{D}/\partial t$ was ignored. Since $|j| = \sigma|E|$, and for sinusoidal field variation $|\partial\bar{D}/\partial t| = \omega\varepsilon|E|$, it follows that this assumption is acceptable provided $\sigma \gg \omega\varepsilon$. This is true for most good conductors for all but the very highest frequencies.

PROBLEMS

P46 (*a*) Prove that the power loss per m^2 of the slab surface shown in Figure 5.3 is given by $P = H_0^2/\sigma\delta$ where H_0 is the rms value of the field at the surface.
(*b*) Calculate the power loss in MW m^{-2} for a slab of steel (resistivity $= 20 \times 10^{-8}\,\Omega\text{m}$, $\mu_r = 500$, $H_0 = 5 \times 10^4\,\text{A m}^{-1}$ and frequency $= 50\,\text{Hz}$). *Hint*: for (*a*) evaluate the average of $(1/\sigma)\int_0^\infty j_x^2(z,t)\,dz$.
(*Answer:* (*b*) $0.35\,\text{MW m}^{-2}$.)

P47 A circular cross-section copper conductor (resistivity $= 0.017 \times 10^{-6}\,\Omega$ m, diameter = 1 cm) is operated at 50 Hz and at 5 kHz. After making the necessary simplifying assumptions, calculate the ratio of the conductor resistance at these two frequencies.
(*Answer:* 2.71.)

P48 If P_h is the power dissipated (in W m^{-2}) in the slab layer $0 < z < h$, and P is the total power loss calculated in Problem P46, prove that $P_h/P = 1 - e^{2h/\delta}$. Calculate the percentage of the total losses developed in the surface layer extending to a depth 2δ.
(*Answer:* 98.2%.)

5.3 LOSSES IN DIELECTRIC AND MAGNETIC MATERIALS

Losses in materials used in electromagnetic applications affect significantly the design of equipment. Indeed, in many cases, the size and rating of equipment is limited by thermal considerations which are ultimately determined by material losses. Losses reduce equipment efficiency and, if cooling is not provided, can cause overheating. The cooling plant required can be very expensive. It is therefore crucial to understand the origin of losses and develop models which permit their calculation. For clarity, dielectric and magnetic materials are studied separately.

All dielectric materials, to a greater or lesser extent, exhibit lossy behaviour. An 'ideal' dielectric material is a useful abstraction but, strictly, it is not found in practice. The readjustment of atomic or other systems inside a dielectric material which follows the application of the field (polarization) results in losses which can be conveniently attributed to friction during polarization. A useful way of looking at this process is to construct a model of a real dielectric, consisting of a capacitor C (ideal component describing energy storage in the electric field) in series with a resistor R (loss element to account for losses) as shown in Figure 5.6(*a*). For a dielectric of reasonable quality R will be very small compared with the magnitude of the capacitor impedance at the frequency of interest ($R \ll 1/\omega C$). Hence the phase angle between the voltage V and the current I will be almost equal to $\pi/2$. Losses cause the phase angle to deviate from $\pi/2$ by a small value δ known as the loss angle.

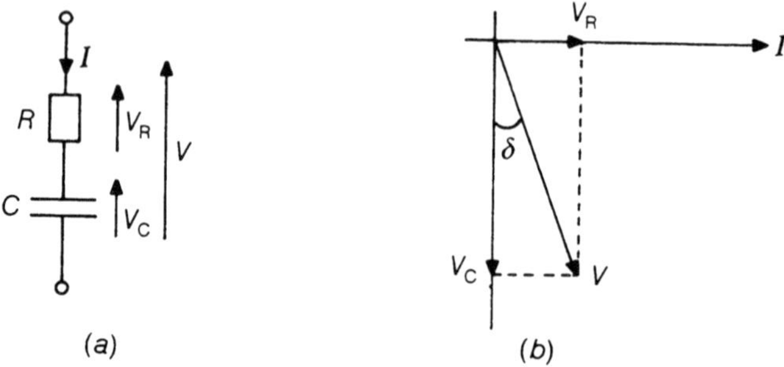

Figure 5.6 (*a*) Equivalent circuit to account for losses and (*b*) loss tangent.

From the phasor diagram of Figure (5.6(*b*)

$$\tan\delta = V_R/V_C = IR/(I \times 1/\omega C) = \omega RC.$$

Hence $R = \tan\delta/\omega C$, where $\tan\delta$ is known as the loss tangent and is a fundamental property of the material (in general it is frequency and temperature dependent). It is evident from Figure 5.6(*a*) that following the application of a step voltage a certain delay is required before steady state is reached inside the dielectric. This phenomenon, known as relaxation, affects also the high-frequency behaviour of the material. When the period of the signal exceeds the time constant RC, the material is able to respond (charge or discharge) as dictated by the field, whilst at high frequencies when $T \ll RC$, the material is unable to respond to fast signal changes. In real materials of complex chemical compositon several polarization processes coexist such as electronic (very fast, lossless, nearly ideal), atomic, molecular, etc. A certain amount of conduction is also inevitable. Hence the equivalent circuit of a real material is as shown in Figure 5.7. The purely resistive path accounts for free-charge conduction (leakage), the purely capacitive path for fast polarization processes (e.g. electronic), and the R–C paths, of which there may be several, account for slow polarization processes. In most applications it is not necessary to use such a complex model (for example, when operation over a restricted frequency range is envisaged) and useful results may be obtained from the simple equivalent circuit shown in Figure 5.6(*a*), where component values are calculated to suit the intended use of the model.

So far losses have been looked at from the circuit point of view. An alternative approach is to use field concepts as shown below.

Let us consider the small block of dielectric material shown in Figure 5.8(*a*), where the two hatched faces are equipotentials with a potential difference V. The equivalent circuit for this block of material is shown in Figure 5.8(*b*), where losses are modelled by adding resistance in parallel with C. This change from the series model shown in Figure 5.6(*a*) may puzzle the reader. It is, however, entirely legitimate. The choice of a parallel or series equivalent circuit is a matter of convenience, although in some cases one or other of the two models may be conceptually more appealing. Naturally, the value of resistance used in the two models is different. It is straightforward for every series circuit to obtain an equivalent parallel circuit and vice versa. From the circuit in Figure 5.8(*b*) and the phasor diagram in Figure 5.8(*c*), $\tan\delta = I_R/I_C = 1/\omega RC$ and the losses are $P = V^2/R = V^2\omega C \tan\delta$. But $V = (E\Delta l)$ and $C = \varepsilon_r\varepsilon_0(\Delta l)^2/\Delta l$, thus yielding

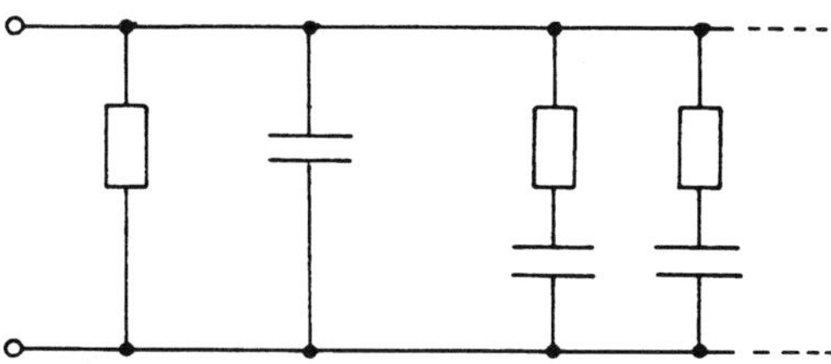

Figure 5.7 A general equivalent circuit for a lossy dielectric.

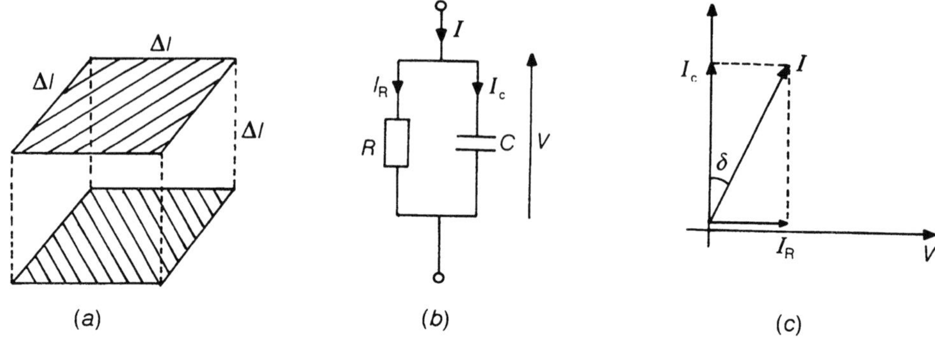

Figure 5.8 Loss characterization using a parallel equivalent circuit.

$P = E^2\varepsilon_r\varepsilon_0\omega \tan\delta(\Delta l)^3$. The losses per unit volume inside the dielectric material are

$$P_l = E^2\varepsilon_r\varepsilon_0\omega \tan\delta \qquad (\mathrm{W\,m^{-3}}) \tag{5.10}$$

where E is the rms value of the electric field. In studying many field problems it is mathematically convenient to incorporate dielectric losses into the dielectric permittivity by making this quantity a complex number (see Problem P49). Alternatively, an equivalent electrical conductivity can be introduced into the formulation. It remains to examine whether the conditions across the boundary between two real dielectrics have to be modified. This problem is best illustrated by an example.

EXAMPLE E30

A parallel-plate capacitor consisting of two layers of real dielectric is shown in Figure E30.

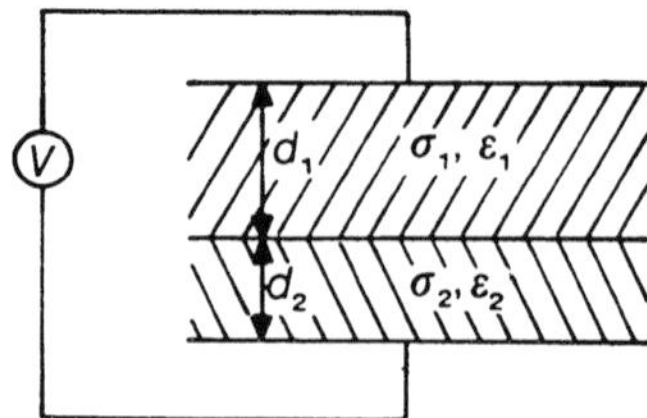

Figure E30

(*a*) Derive an expression for the ratio of the values of the field in the two dielectrics E_1/E_2 in steady state, for a dc and for a high-frequency applied voltage V.
(*b*) For the case of a dc applied voltage, obtain an expression for the surface free-charge density σ_f at the boundary between the two dielectrics.

Solution

(*a*) The reader must resist the temptation to use equation (2.22) which was derived on the basis that there were no free charges in the dielectrics. The current density anywhere inside the dielectrics is $j = \sigma E + \partial D/\partial t$, the first term representing losses (current through R in the equivalent circuit of Figure 5.8(*b*)), and the second the displacement current (current through C in Figure 5.8(*b*)). Under steady state dc conditions, the second term is zero, and continuity of current demands $\sigma_1 E_1 = \sigma_2 E_2$ where E_1, E_2 are the electric strengths in the two dielectric layers. Hence it follows that $E_1/E_2 = \sigma_1/\sigma_2$. At very high freuencies the second term predominates over the first (since for a sinusoidal current $|\partial D/\partial t| = \omega\varepsilon E$) and $E_1/E_2 = \varepsilon_1/\varepsilon_2$ in accordance with equation (2.22). Very high frequencies in this context means frequencies such that $\omega\varepsilon \gg \sigma$. Clearly, at intermediate frequencies both σ and ε affect the electric field distribution. As it is unlikely that $\sigma_1/\sigma_2 = \varepsilon_1/\varepsilon_2$, the two dielectric layers will be stressed differently at low and at high frequencies. Commercially available capacitors have a different voltage rating for low- and high-frequencies operation. It should be noted that the specification in this example of steady state conditions for a dc applied voltage was necessary, since during the transient (charging up of the capacitor) the displacement term must be substantial.

(*b*) The free-charge density on the boundary can be easily obtained by applying Gauss's law as was shown in section 2.5. It is then found that $D_1 - D_2 = \sigma_f$. Substituting

$$\sigma_f = D_1 - D_2 = \varepsilon_1 E_1 - \varepsilon_2 \frac{\sigma_1}{\sigma_2} E_1 = E_1 \frac{\varepsilon_1\sigma_2 - \varepsilon_2\sigma_1}{\sigma_2}.$$

But

$$V = E_1 d_1 + E_2 d_2 = \frac{d_1\sigma_2 + d_2\sigma_1}{\sigma_2} E_1.$$

Combining these two expressions gives

$$\sigma_f = \frac{\varepsilon_1\sigma_2 - \varepsilon_2\sigma_1}{d_1\sigma_2 + d_2\sigma_1} V \qquad (\mathrm{C\,m^{-2}}).$$

Let us now direct attention to losses in magnetic materials. Much has already been said about losses in connection with eddy currents. There are, however, two distinct mechanisms responsible for losses in magnetic materials, eddy current and hysteresis losses, known collectively as iron losses. Eddy currents are due to induced voltages caused by changing flux (Faraday's law). It is easy to obtain a feel for these losses. Let V_{ind} be the induced voltage and R the resistance of an imaginary loop in the magnetic material. Eddy losses are then equal to V_{ind}^2/R and since $V_{ind} \sim fB$ it follows that these losses are proportional to f^2. Eddy current losses may be reduced by proper design (laminations). The other loss mechanism is due to hysteresis. The process of alignment of magnetic domains to the applied field causes 'frictional' losses which can be calculated from the B–H curve of the material. During magnetization (Ob in Figure 5.9) an amount of energy proportional to the area OabedO is supplied by the source. Reduction of the supply current to zero brings the

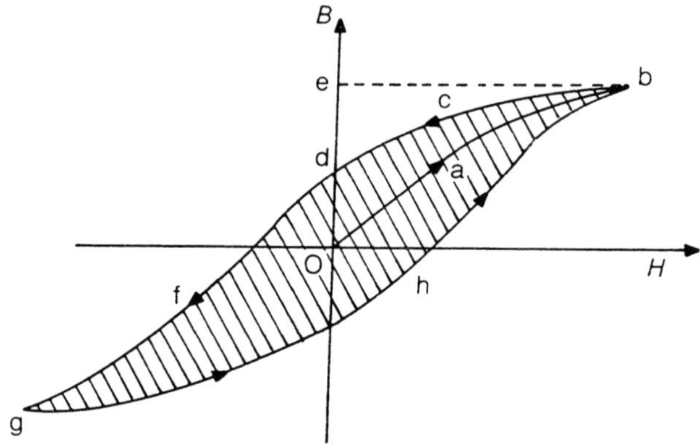

Figure 5.9 The hatched area inside the hysteresis loop is a measure of hysteresis losses.

state of magnetization of the material to point d and an amount of energy proportional to the area bedcb is returned to the supply. Hence the area OabdO represents losses in the material during this process. By similar arguments it can be shown that for a complete hysteresis loop, the hatched area is a measure of the hysteresis losses. This area represents the losses for a complete cycle of magnetization (or supply current). The energy loss in one second (hysteresis power loss) is then proportional to the hatched area and the number of cycles in one second, i.e. the frequency. Hysteresis losses can be reduced by selecting materials with a thin hysteresis loop.

Calculation of iron losses under practical conditions is difficult especially when strong non-linearities are present.

REMARKS

In real dielectrics, losses (conduction and dielectric) are always present. In mathematical models losses are represented by current components in phase with the field (conduction current σE). There are also current components $\pi/2$ out of phase with the field (displacement current $\partial D/\partial t$). In these models σ and ε represent a whole host of physical mechanisms which will not be described here. Although dielectric losses are presented in this secion as a problem, there are applications where they can be exploited to heat materials (e.g. microwave cooking of foods). Losses in magnetic materials are very important as they affect the efficiency of many electromagnetic devices. Substantial materials research effort is put into developing less lossy materials with the correct mechanical properties for use in electrical and electronic applications.

PROBLEMS

P49 A unit volume of real dielectric material is represented by the equivalent circuit shown in Figure P49. An alternative is to represent the material by. an

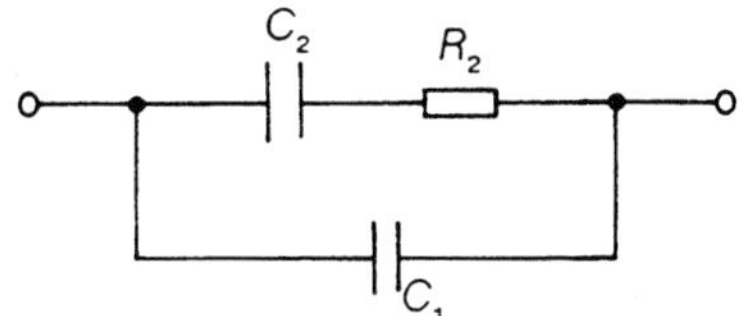

Figure P49

equivalent capacitor of admittance $j\omega\varepsilon$ where ε is a complex number $\varepsilon = \varepsilon' - j\varepsilon''$. Calculate $\varepsilon', \varepsilon''$ in terms of the components shown in Figure P49. Sketch the variaton of ε' and $\tan\delta = \varepsilon''/\varepsilon'$ with frequency.
Hint: obtain the total admittance of the circuit in Figure P49 and equate it with $j\omega\varepsilon$.
(*Answer*: $\varepsilon' = C_1 + C_2/[1 + (\omega R_2 C_2)^2]$, $\varepsilon'' = \omega R_2 C_2^2/[1 + (\omega R_2 C_2)^2]$.)

P50 A coaxial cable with inner radius $r_1 = 1$ cm and outer radius $r_2 = 3$ cm is filled with a dielectric $\varepsilon_r = 3$, $\tan\delta = 0.01$. A voltage of frequency 1 kHz and rms value 10 kV is applied across the two conductors. Calculate the total power loss per metre length of the cable.
(*Answer*: 0.95 W m^{-1}.)

5.4 DISTRIBUTED PARAMETER CIRCUITS—TRANSMISSION LINES

A wide range of electromagnetic devices can be studied by constructing models consisting of lumped circuit elements, i.e. capacitors, inductors and resistors. This practice appears to be justifiable when circuits are built using components obtained from stores, where each component comes with a declaration of its function. Even in this case, as shown in section 5.1, difficulties arise as the frequency varies, since the nature of each component can change substantially. At very high frequencies or when fast pulses are present, a fundamental rethink of the nature of components is necessary. There are also systems, such as power and telecommunication transmission lines, where it is not immediately obvious where the capacitance, inductance or resistance reside. All three are mixed together and the practice of separating (lumping) each component is called into question. The purpose of this section is to clarify these issues and establish rules for constructing adequate models for particular systems and problems.

Let us consider a long transmission line consisting of two parallel plates as shown in Figure 5.10(*a*). It is assumed that the separation h is small enough to justify the assumption that electric and magnetic fields do not vary substantially along the z-direction. Exchange of information between the two plates along the z-direction is instantaneous. These assumptions are acceptable provided the propagation time between the two plates h/v_0 is insignificant compared with the period of variation of the signals, i.e. $h/v_0 \ll T$. The speed of propagation v_0 is equal to 3×10^8 m s^{-1} when the medium surrounding the plates is air. The width of the line is assumed large enough (say $w > 3h$) so that most of the field is concentrated between the plates. This is not an assumption of fundamental nature—it is simply necessary to permit easy calculation of the parameters (inductance, capacitance) of the line. No restriction is

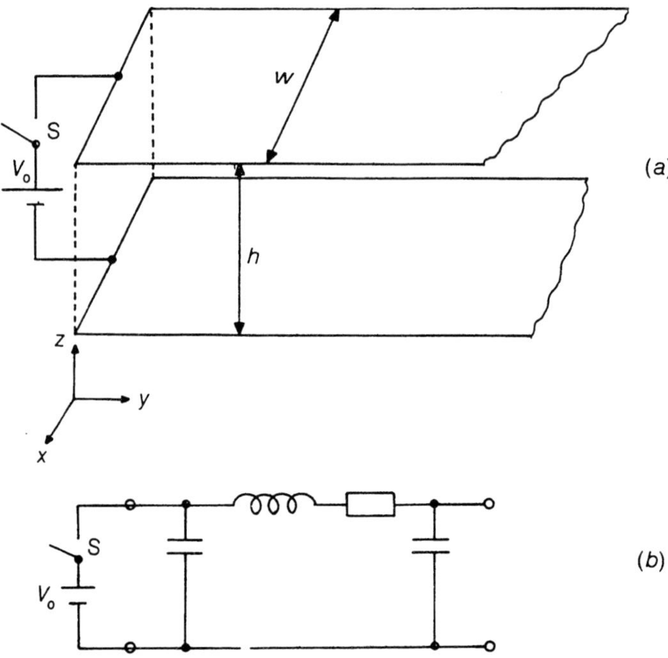

Figure 5.10 (*a*) Parallel plate transmission line and (*b*) lumped parameter equivalent circuit.

placed on the length of the line (dimension along y) apart from requiring that the line is long. The line is energized upon closing switch S whereupon the voltage V_0 is impressed on the line. In order to study the response of the line to the application of the voltage V_0, it is necessary to construct a circuit model. One approach would be to calculate the total line capacitance (between the parallel plates), the total inductance (due to the parallel plate conductors) and whatever ohmic or other losses are associated with the line, and construct the model of the line shown in Figure 5.10(*b*), where the total capacitance has been shared equally between the two ends of the line to emphasize its symmetry. This is essentially a lumped equivalent circuit for the line. It is obvious that following energization, a response will immediately appear at the output. This cannot be right especially when the line is long. A network which is physically long cannot be modelled in this way. However, if a slowly varying signal is applied at the input, then the fact that, strictly, the output should appear after a short time interval may not be of major significance. In this case (when line travel time is very much less than the period of the signal) this simple equivalent circuit can give acceptable results. It is when the period of the signal (or other time interval characteristic of the rate of change of the signal) is comparable or smaller to the line travel time, that the circuit model in Figure 5.10(*b*) becomes inadequate. Under these circumstances the distributed nature of the line must be taken into account.

Let us then abandon the circuit in Figure 5.10(*b*) and examine from first principles the line energization problem shown in 5.10(*a*). Following the closure of S a disturbance will propagate down the line with a velocity, as yet unknown, v_0.

After a time interval Δt, the disturbance would have penetrated a length $\Delta l = v_0 \Delta t$ down the line, and this segment of the line would have charged to a voltage V_0. The charge deposited on the line would be $\Delta q = C_l \Delta l V_0$, where C_l is the capacitance per unit length of the line. The line current can then be found, $I = \Delta q/\Delta t = V_0 C_l(\Delta l/\Delta t) = V_0 C_l v_0$. The voltage on the line can also be calculated from Faraday's law: $V_0 = \Delta\phi/\Delta t = \Delta(L_l \Delta l I)/\Delta t$ where L_l is the inductance per unit length. Hence, substituting for the current, $V_0 = C_l L_l V_0 v_0^2$ and the velocity of propagation on the line is

$$v_0 = \frac{1}{\sqrt{L_l C_l}} \qquad (\mathrm{m\,s^{-1}}). \qquad (5.11)$$

It should be noted that v_0 depends entirely on the line parameters. Substituting into the expression for the current gives

$$I = V_0 C_l \frac{1}{\sqrt{L_l C_l}} = \frac{V_0}{\sqrt{L_l/C_l}}.$$

The expression in the denominator has the dimensions of impedance and it is known as the characteristic impedance of the line:

$$Z = \sqrt{L_l/C_l} \qquad (\Omega). \qquad (5.12)$$

To summarize, following the closure of S a disturbance will propagate down the line with a velocity given by (5.11). The disturbance voltage is V_0 and the associated current is $I = V_0/Z$. At this stage it matters not how the line is terminated. The current is determined by the characteristic impedance Z. Until the disturbance has reached the termination and a return disturbance has travelled back to the source the current near the source remains constant at $I = V_0/Z$. Energy will be supplied into the line at the rate $P = VI = V_0^2/Z$ (in units of W). It is straightforward, based on the assumptions already mentioned, to calculate C_l and L_l. Since fringing is neglected

$$C_l = \frac{\varepsilon \Delta l w/h}{\Delta l} = \frac{\varepsilon w}{h} \qquad (\mathrm{F\,m^{-1}}).$$

Applying Ampere's law on curve C shown in Figure 5.11 and neglecting contri-

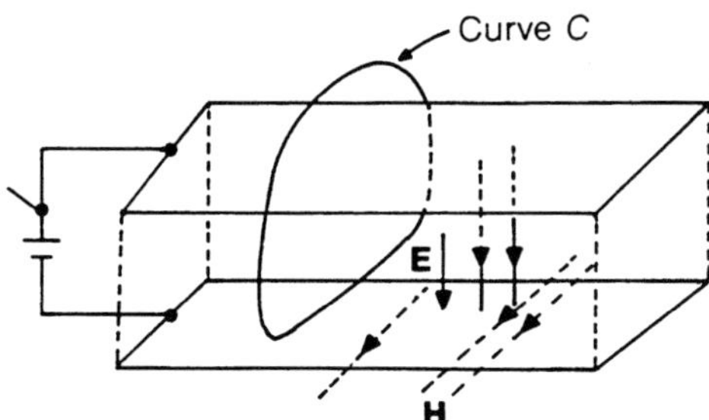

Figure 5.11 Electric and magnetic field associated with line.

butions from the space outside the plates gives $Hw = I$; hence $\phi = (\mu H)h\Delta l = \mu(h/w)\Delta l I$ and

$$L_l = \frac{\phi/I}{\Delta l} = \mu h/w \qquad (\mathrm{H\,m^{-1}}).$$

The approximations inherent in these calculations affect the accuracy of determination of C_l and L_l, but not the development of the physical picture of propagation of the disturbance. Substituting the values of C_l and L_l into (5.11) and (5.12) gives

$$v_0 = 1/\sqrt{\mu\varepsilon} \tag{5.13}$$

$$Z = (h/w)\sqrt{\mu/\varepsilon}. \tag{5.14}$$

If the insulation medium for the line is air, then $v_0 = c = 3 \times 10^8\,\mathrm{m\,s^{-1}}$ and $Z = 377\,\mathrm{h\,w^{-1}}$ (expressed in Ω).

Let us now construct the picture of propagation from the field point of view. The propagation of the voltage and current disturbance down the line will generate associated electric and magnetic fields as shown in Figure 5.11. The magnitude of these field components is

$$E = V_0/h \qquad \text{and} \qquad H = I/w = V_0/Zw.$$

The quantity $S = EH = V_0^2/(hwZ)$ represents the power per unit cross-sectional area propagating along the line, as comparison with the circuit formula will confirm. It can be shown that, in general, the power flux can be designated by a vector **S**, known as Poynting's vector, such that

$$\mathbf{S} = \mathbf{E} \times \mathbf{H} \qquad (\mathrm{W\,m^{-2}}). \tag{5.15}$$

The expression found for the example of Figure 5.10(*a*) is a special case of (5.15) when **E** and **H** are perpendicular to each other. It should be noted that (5.15) gives not only the magnitude but also the direction of the flux.

REMARKS

The reader may be intrigued by the eventual outcome of the energization process, the initial phase of which was described in this section. What, for example, will happen if the line is terminated by an open circuit? Clearly, when the disturbance (voltage V_0 and current V_0/Z) arrives at the open circuit termination, the current at the termination must be zero (since this is the nature of the boundary condition imposed by an open circuit). Hence another disturbance must travel towards the source such that the current is forced to zero. This disturbance will be superimposed on the original disturbance (see Problem P52). Only when the impedance of the termination is equal to the characteristic impedance of the line are there no reflections. The line then is said to be matched. In general, the process of energizing the line consists of a succession of reflected disturbances until steady state is reached. For the case of open circuit termination, the steady state is $I = 0$ and $V = V_0$ everywhere on the line. If there are no losses on the line (no damping) the steady state consists of oscillatory voltage and current waveforms with average value V_0 and zero respectively.

It is worth emphasizing the importance of the characteristic impedance (also referred to as the surge impedance) Z of the line, especially in high-frequency and pulse applications. Indeed, during the initial energization phase of the circuit shown in Figure 5.10(*a*), Z is effectively the input impedance of the line as seen by the source. When a coaxial cable is specified for instrumentation purposes (50 Ω) or for a TV aerial connection (75 Ω), what is actually specified is the characteristic impedance of the cable as defined in (5.12).

If losses are included, Z also has a reactive part.

Analytical results are available to study the steady state response of the line to a sinusoidal excitation. These results can be derived by constructing a model of the entire length of the line by cascading a large number of segments, such as that shown in Figure 5.10(*b*), each representing a short line segment Δl such that $\Delta l/v_0 \ll T$. In this way lumped circuits can be used to study, approximately, distributed systems. It is also worth emphasizing that, although a transmission line is the most common example of a distributed system, there are many other systems which, under certain circumstances, behave as distributed networks. An example of such a system is a transformer winding. Under low-frequency conditions (50 or 60 Hz), it can be adequately represented by a lumped inductance. However, under pulse conditions (as for, example, when subject to a lightning disturbance) it behaves as a distributed network not unlike a transmission line.

PROBLEMS

P51 Determine which of the following problems should be solved using distributed parameter equivalent circuits:
(i) overhead power line, 100 km long, 50 Hz excitation;
(ii) overhead power line, 100 km long, subject to a lightning impulse (1 μs risetime);
(iii) signal propagation (100 MHz) on a 1.5 m long coaxial cable;
(iv) propagation of a 5 ns risetime pulse between two adjacent parallel tracks of a printed circuit board 30 cm in length.
(*Answer:* all except (i) but why?)

P52 Show that a voltage pulse propagating on a lossless transmission line of characteristic impedance Z, and transit time τ, doubles in magnitude upon reflection from an open circuit termination. *Hint:* impose energy conservation over a period 2τ. At $t = 2\tau$ all energy is stored in the electric field.

P53 A parallel-plate air-insulated line (height = 1 cm, width = 2 cm) is used in a pulsed power application. To avoid electrical breakdown, the electric field must not exceed 10 kV cm^{-1}. All edge effects are neglected. Calculate the maximum amount of power that can be transmitted using this line.
(*Answer:* 0.53 MW.)

P54 A parallel-plate line is insulated using a plastic material ($\varepsilon_r = 3$, $\mu_r = 1$) and its cross-sectional dimensions and length are $h = 5$ mm, $w = 15$ mm, $l = 30$ cm. A step voltage source (voltage = 10 V, internal resistance = 100 Ω) is connected to one end of the line, whilst the other end is open circuit. Calculate the time interval required for the disturbance to reach the open circuit end, and the value of the voltage measured there just after the arrival of the disturbance.
(*Answer:* 1.73 ns, 11.14 V.)

CHAPTER 6

Maxwell's Equations

The reader will be relieved to know that a very substantial ground has already been covered in deriving the fundamental laws of electromagnetism. On several occasions expressions were given simply as experimental laws (e.g. Coulomb force, Biot–Savart law) and a multitude of other expressions were derived to cover a range of phenomena. The impression may have been formed that electromagnetic theory is a collection of disjointed equations each chosen to fit particular situations. This is not the case. Every well developed physical theory is based on a very small number of basic premises out of which, by mathematical manipulation, a whole range of useful expressions is obtained. The fewer the basic premises, the more abstract and mathematically involved is the derivation of the useful practical expressions. Ultimately, everything that has been stated so far can be derived from a single basic premise (the minimum possible in intellectual work, otherwise it appears that one is led to circular arguments). The point at which each author of a textbook chooses to start, is largely a matter of taste and of the intended purpose of the book. Many authors start from the so-called Maxwell equations. Others follow a more gradual, phenomenological treatment. The latter approach is used in this book, as it delays the introduction of powerful but difficult to use mathematical tools, until familiarity with the basic concepts and applications has been established. Maxwell's equations are, however, an elegant codification of electromagnetism and they are formally introduced in this section. This will lead to more advanced work in the final part of the book.

The basic understanding of fields is embodied in the set of equations given below, and which are known collectively as Maxwell's equations in integral form:

$$\text{Ampere's law} \quad \oint_C \mathbf{H}\cdot d\mathbf{l} = \int_S \left(\mathbf{j}_f + \frac{\partial \mathbf{D}}{\partial t}\right)\cdot d\mathbf{s} \qquad (4.3)$$

$$\text{Faraday's law} \quad \oint_C \mathbf{E}\cdot d\mathbf{l} = -\int_S \frac{\partial \mathbf{B}}{\partial t}\cdot d\mathbf{s} \qquad (4.4)$$

$$\text{Gauss's law} \int_S \mathbf{D}\cdot d\mathbf{s} = \int_V \rho_f \, dv \qquad (2.20)$$

$$\text{Flux conservation law} \int_S \mathbf{B} \cdot \mathrm{d}\mathbf{s} = 0. \tag{6.1}$$

These are supplemented by the constitutive equations:

$$\mathbf{j}_\mathrm{f} = \sigma \mathbf{E} \qquad \text{Ohm's law} \tag{3.3}$$

$$\mathbf{D} = \varepsilon_\mathrm{r}\varepsilon_0 \mathbf{E} \tag{2.19}$$

$$\mathbf{B} = \mu_\mathrm{r}\mu_0 \mathbf{H} \tag{3.8}$$

the force equation:

$$\mathbf{F} = q(\mathbf{E} + \mathbf{V} \times \mathbf{B}) \tag{3.5}$$

and the continuity equation:

$$\int_S \mathbf{j}_\mathrm{f} \cdot \mathrm{d}\mathbf{s} = -\frac{\partial Q}{\partial t}. \tag{6.2}$$

This last equation simply says that the total current emerging from a closed surface is at the expense of the charge Q enclosed by it (charge conservation).

Armed with these expressions the tenacious reader can, in principle, tackle any problem in electromagnetism. One feature of these expressions is that they describe relationships over a large part of the system studied (integral laws). For instance, Gauss's law relates the electric field over a closed surface to charges in the volume enclosed by this surface. This feature can sometimes lead to difficulties. It has already been pointed out how important the selection of this surface is in order to obtain results easily. In cases where the field lacks any obvious symmetry, these laws can only be exploited by resorting to numerical schemes. Other circumstances where the integral laws are difficult to exploit are when the material in which the field is established is inhomogeneous or when its properties are strongly non-linear. For these reasons, another form of these laws is useful, where the surface, curve or volume, on which the integral laws are applied, is reduced so that, in the limit, it collapses into a single point. At this limit, Maxwell's equations take their 'differential' or point form, and relate the value of electromagnetic quantities at a point.

In approaching this limit, it will be necessary to deal with two different types of calculation. Firstly, calculation of quantities of the type

$$\lim_{\Delta v \to 0} \frac{\int_S \mathbf{G} \cdot \mathrm{d}\mathbf{s}}{\Delta v}$$

will be necessary, where Δv is the volume enclosed by surface S, and $\mathbf{G}$ is any vector quantity such as $\mathbf{D}$ or $\mathbf{B}$. The expression above can be calculated in the limit and it is known as the divergence (div) of the vector $\mathbf{G}$, written as div$\mathbf{G}$ or symbolically as $\nabla \cdot \mathbf{G}$:

$$\nabla \cdot \mathbf{G} = \lim_{\Delta v \to 0} \frac{\int_S \mathbf{G} \cdot \mathrm{d}\mathbf{s}}{\Delta v}.$$

The divergence of a vector is a scalar and in a cartesian coordinate system it is given by

$$\nabla\cdot\mathbf{G}=\frac{\partial G_x}{\partial x}+\frac{\partial G_y}{\partial y}+\frac{\partial G_z}{\partial z}. \tag{6.3}$$

Appropriate expressions for $\nabla\cdot\mathbf{G}$ can be found in other coordinate systems (see Appendix B).

Secondly, calculation of quantities of the type

$$\lim_{\Delta s\to 0}\frac{\oint_C \mathbf{G}\cdot d\mathbf{l}}{\Delta s}$$

will be necessary, where ΔS is a surface limited by the curve C. $\mathbf{G}$ is any vector quantity such as $\mathbf{E}$ or $\mathbf{H}$. This expression can be calculated in the limit and it is called the curl of the vector $\mathbf{G}$, written as curl $\mathbf{G}$ or symbolically as

$$\nabla\times\mathbf{G}=\lim_{\Delta s\to 0}\frac{\int_C \mathbf{G}\cdot d\mathbf{l}}{\Delta s}.$$

The curl of a vector is itself a vector and in a cartesian coordinate system is given by the expression

$$\nabla\times\mathbf{G}=\left(\frac{\partial G_z}{\partial y}-\frac{\partial G_y}{\partial x}\right)\hat{\mathbf{x}}+\left(\frac{\partial G_x}{\partial z}-\frac{\partial G_z}{\partial x}\right)\hat{\mathbf{y}}+\left(\frac{\partial G_y}{\partial x}-\frac{\partial G_x}{\partial y}\right)\hat{\mathbf{z}} \tag{6.4}$$

where the expressions in brackets are the x, y, z, components of the vector. Appropriate expressions are available for $\nabla\times\mathbf{G}$ in other coordinate systems. As long as the functional dependence of $\mathbf{G}$ on coordinates (x, y, z) is known its div and curl can be obtained from (6.3) and (6.4) respectively. Let us now return to Maxwell's equations and proceed to cast them in differential form starting with Ampere's law.

A small surface ΔS is chosen perpendicular to $\hat{\mathbf{z}}$ and of an area equal to $\Delta x\Delta y$. Then

$$\oint\mathbf{H}\cdot d\mathbf{l}=\left(j_z+\frac{\partial D_z}{\partial t}\right)\Delta x\Delta y.$$

$$j_z\to\frac{\partial D_z}{\partial t}=\lim_{\Delta x\Delta y\to 0}\frac{\oint\mathbf{H}\cdot d\mathbf{l}}{\Delta x\Delta y}=(\nabla\times\mathbf{H})_z$$

or, in more compact form,

$$\nabla\times\mathbf{H}=\mathbf{j}_{\mathrm{f}}+\frac{\partial\mathbf{D}}{\partial t}. \tag{6.5}$$

Faraday's law is similarly treated to give

$$\nabla\times\mathbf{E}=-\frac{\partial\mathbf{B}}{\partial t}. \tag{6.6}$$

Next, Gauss's law can be written as

$$\lim_{\Delta v \to 0} \frac{\int_S \mathbf{D}\cdot \mathrm{d}\mathbf{s}}{\Delta v} = \lim_{\Delta v \to 0} \frac{\int_{\Delta v} \rho_{\mathrm{f}}\, \mathrm{d}v}{\Delta v},$$

where Δv is the volume enclosed by surface S. The expression on the left-hand side is simply div$\mathbf{D}$ and the right-hand side is the charge density. Hence,

$$\nabla\cdot\mathbf{D} = \rho_{\mathrm{f}}. \tag{6.7}$$

Similarly, the law of flux conservation gives

$$\nabla\cdot\mathbf{B} = 0. \tag{6.8}$$

Equations (6.5) to (6.8) constitute Maxwell's equations in differential form. The continuity equations can be similarly expressed as

$$\nabla\cdot\mathbf{j}_{\mathrm{f}} = -\frac{\partial \rho_{\mathrm{f}}}{\partial t}. \tag{6.9}$$

It has been shown in section 2.3 that the electric field due to static charges can be obtained from the electric potential using (2.15). This expression can be put into a more compact form by defining a vector called the gradient of V and written as grad$V$, or symbolically as ∇V, such that

$$\nabla V = \frac{\partial V}{\partial x}\hat{\mathbf{x}} + \frac{\partial V}{\partial y}\hat{\mathbf{y}} + \frac{\partial V}{\partial z}\hat{\mathbf{z}}. \tag{6.10}$$

Hence $\mathbf{E}_{\mathrm{es}} = -\nabla V$, where the subscript es stands for electrostatic component. The potential for any general distribution of charge is given by

$$V = \int_V \frac{\rho_v\, \mathrm{d}v}{4\pi\varepsilon r}. \tag{2.17}$$

However, as has already been pointed out, electric fields can also be generated by changing magnetic fields (Faraday's law). Therefore there are in general two components to the electric field, the electrostatic component given above and obtained from the electrostatic potential V, and an electromagnetic component derived from Faraday's law. It is convenient to define an additional potential function which will furnish not only the electromagnetic component of the electric field, but also the magnetic field. This potential is a vector $\mathbf{A}$, known as the vector potential to distinguish it from V which is referred to as the scalar potential. It can be calculated in analogy with (2.17) but with the current density replacing the charge density:

$$\mathbf{A} = \int_V \frac{\mu \mathbf{j}\, \mathrm{d}v}{4\pi r}. \tag{6.11}$$

The electromagnetic component of the electric field is then $\mathbf{E}_{\mathrm{em}} = -\partial\mathbf{A}/\partial t$, and the magnetic field is $\mathbf{B} = \nabla \times \mathbf{A}$. To summarize, if the charge and current density distributions are known, the scalar and vector potentials can be obtained from (2.17) and (6.11) and the electric and magnetic fields are then calculated using:

$$\mathbf{E} = -\nabla V - \partial \mathbf{A}/\partial t \tag{6.12}$$

$$\mathbf{B} = \nabla \times \mathbf{A}. \tag{6.13}$$

In field computations the procedure described above is normally adopted.

There are, however, many problems where the charge and current densities are not known in advance. Instead, general constraints are placed on the system regarding potential values in sections of the problem (boundary conditions) or field values at the moment of excitation of the system (initial conditions). It is then necessary, using Maxwell's equations, boundary and initial conditions, to obtain a solution for the potentials. This is, in general, a complicated procedure, and in most cases requires the use of advanced analytical and numerical techniques. Let us examine some of the constraints imposed on potentials.

In electrostatic problems $\mathbf{E} = -\nabla V$; hence substituting in (6.7) gives

$$\rho_f = \nabla \cdot \mathbf{D} = \varepsilon \nabla \cdot \mathbf{E} = -\varepsilon \nabla \cdot (\nabla V).$$

The divergence of the gradient of a scalar quantity has a special symbol in mathematics ∇^2 (called del squared) and it can be shown that in cartesian coordinates

$$\nabla^2 V = \frac{\partial^2 V}{\partial x^2} + \frac{\partial^2 V}{\partial y^2} + \frac{\partial^2 V}{\partial z^2}.$$

Hence for the scalar potential

$$\nabla^2 V = -\rho_f/\varepsilon. \tag{6.14}$$

This is known as Poisson's equation. In regions where free charges are not present ($\rho_f = 0$), it reduces to Laplace's equation:

$$\nabla^2 V = 0. \tag{6.15}$$

Similar expressions can be obtained for the vector potential $\mathbf{A}$. From (6.5), assuming slow field variations and high conductivity ($\partial D/\partial t \ll j_f$),

$$\nabla \times \mathbf{H} = \mathbf{j}_f, \text{ or } \quad \nabla \times \mathbf{B} = \mu_0 \mathbf{j}_f.$$

Hence, using standard vector identities,

$$\nabla \times \mathbf{B} = \nabla \times (\nabla \times \mathbf{A}) = -\nabla^2 \mathbf{A} + \nabla(\nabla \cdot \mathbf{A}) = \mu_0 \mathbf{j}_f.$$

But $\mathbf{A}$ can be chosen so that $\nabla \cdot \mathbf{A} = 0$; hence

$$\nabla^2 \mathbf{A} = -\mu_0 \mathbf{j}_f. \tag{6.16}$$

The x-component of (6.16) is $\nabla^2(\mathbf{A}_x \hat{\mathbf{x}}) = -\mu_0 j_x \hat{\mathbf{x}}$, and similar expressions are obtained for the y- and z-components. Equation (6.16) is useful in studying, for example, eddy current problems. Solutions to many other problems can be found using these equations, provided a certain standard in mathematical manipulation has been reached.

EXAMPLE E31

Obtain a solution to Problem P9 using the differential form of Maxwell's equations.

Solution

This is a case where the electric field varies along the x-direction only, hence $\mathrm{div}\mathbf{D} = \rho_f$ reduces to $\mathrm{d}E_x/\mathrm{d}x = \rho_f/\varepsilon$. Integrating this equation for $0 < x < a$ gives $E_x = \rho_v x/\varepsilon + C$, where C is a constant. The boundary conditions dictate that for $x = 0$, $E = 0$ hence $C = -\rho_v a/\varepsilon$ and $E_x = (\rho_v/\varepsilon)(x - a)$. Similarly, for $-a < x < 0$, $E_x = -\rho_v x/\varepsilon + C'$ and the boundary condition at $x = 0$ gives $C' = -\rho_v a/\varepsilon$, hence $E_x = -(\rho_v/\varepsilon)(x + a)$. An alternative approach is to calculate the scalar potential directly by solving Poisson's equation. Since V varies only with x, $\nabla^2 V = \mathrm{d}^2 v/\mathrm{d}x^2$. Hence

$$\frac{\mathrm{d}^2 v}{\mathrm{d}x^2} = -\rho/\varepsilon.$$

Integrating once gives $\mathrm{d}V/\mathrm{d}x = -\rho x/\varepsilon + C_1$ and once more gives $V = -\rho x^2/2\varepsilon + C_1 x + C_2$ where C_1, C_2 are constants to be determined from the boundary conditions. Setting the reference $V = 0$ at $x = 0$, dictates that $C_2 = 0$. The electric field at $x = a$ should also be zero, and since $E = -\mathrm{d}V/\mathrm{d}x$ it follows that $-\rho_v a/\varepsilon + C_1 = 0$. Hence $C_1 = \rho_v a/\varepsilon$ and $V(x) = -(\rho_v/\varepsilon)(x^2/2 - ax)$ (for $0 < x < a$).

The electric field can now be found:

$$E = -\nabla V = -\frac{\mathrm{d}V}{\mathrm{d}x} = \rho_v(x - a)/\varepsilon.$$

The reader should compare these results with those obtained in Problem P9 using the integral form of Gauss's law.

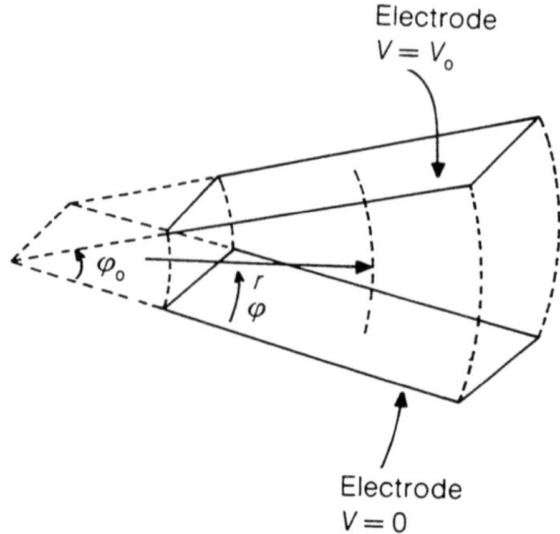

Figure E32

EXAMPLE E32

A potential difference V_0 is applied across two electrodes arranged as shown in Figure E32. All edge effects are neglected. Obtain an expression for the scalar potential and for the electric field, valid anywhere between the two electrodes.

Solution

The problem has cylindrical symmetry, hence it is profitable to use a cylindrical system of coordinates. The potential varies only with the angle φ, hence $\nabla^2 V = 0$ reduces to

$$\frac{1}{r^2}\frac{\partial^2 V}{\partial \varphi^2} = 0.$$

(All other terms in $\nabla^2 V$ are zero; see Appendix B for $\nabla^2 V$ in cylindrical coordinates.)

Integrating the expression above gives $V(\varphi) = C_1\varphi + C_2$. The boundary conditions are $V = 0$ at $\varphi = 0$ and $V = V_0$ at $\varphi = \varphi_0$, hence $C_2 = 0$ and $C_1 = V_0/\varphi_0$. The scalar potential between the electrodes is then $V(\varphi) = V_0\varphi/\varphi_0$. The electric field is $\mathbf{E} = -\nabla V$ and in this case only $\mathbf{E}\varphi$ has a non-zero value. From Appendix B,

$$E_\varphi = \frac{1}{r}\frac{\partial V}{\partial \varphi} = \frac{1}{r}\frac{V_0}{\varphi_0}.$$

EXAMPLE E33

(*a*) Obtain an expression for the vector potential due to a long straight conductor.
(*b*) Using the vector potential found in (*a*), outline a procedure for calculating the magnetic field due to a conductor of an arbitrary cross-section.

Solution

(*a*) This is a repetition of Example E18, using the vector potential formalism. Keeping the same notation as in E18, the component of **A** at P due to the current element at O is

$$\mathrm{d}A_z = \frac{\mu_0 I\,\mathrm{d}l}{4\pi R} = \frac{\mu_0 I\,\mathrm{d}l}{4\pi\sqrt{d^2 + l^2}}.$$

If the wire extends from $-L$ to $+L$, then (using tables of integrals)

$$A_z = \int_{-L}^{L} \frac{\mu_0 I\,\mathrm{d}l}{4\pi\sqrt{d^2 + l^2}} = \frac{\mu_0 I}{4\pi}\ln\frac{\sqrt{1 + (d^2/L^2)} + 1}{\sqrt{1 + (d^2/L^2)} - 1}.$$

For a very long wire ($L \rightarrow \infty$), $A_z = (\mu_0 I/2\pi)\,(\ln 2L - \ln d)$. The first term tends to a

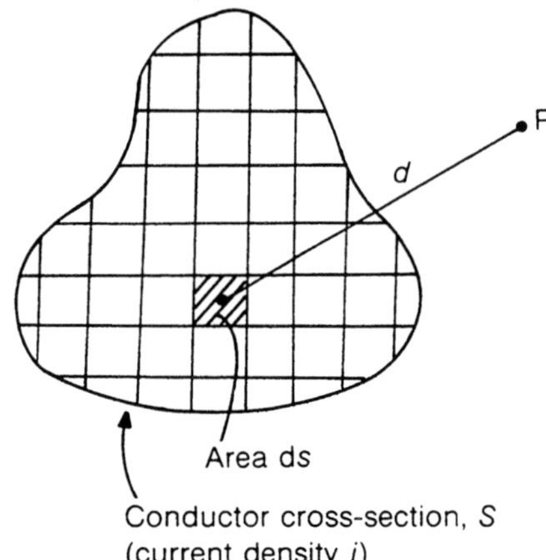

Figure E33

large constant value. Since **E** and **B** are obtained from **A** through differentiation, the large constant term makes no contribution to the field values. Hence for calculating field components

$$A_z = -\frac{\mu_0 I}{2\pi} \ln d.$$

All other components (A_x, A_y) are zero.
(*b*) Let us assume that the current is distributed uniformly (low-frequency case) over the cross-section of the conductor shown schematically in Figure E33. It can be imagined that the conductor is divided into small sections as shown, such that the field due to each section can be found using either the techniques introduced in Example E18 or the vector potential calculated in (*a*). In both cases the combination of all such contributions yields the field due to the entire conductor. In the former case where **B** due to each segment is calculated, the combination of all contributions is a true vector operation since each contribution is in a different direction. In contrast, in the latter case, the contribution to the vector potential from each segment is in the same z-direction. Hence combining all contributions is a straightforward matter. Thus

$$\mathrm{d}A_z = -\mu_0 \frac{j\,\mathrm{d}s}{2\pi} \ln d$$

and

$$A_z = -\frac{\mu_0 j}{2\pi} \int \ln d\,\mathrm{d}s = -\frac{\mu_0 I}{2\pi}\left(\frac{1}{S}\int \ln d\,\mathrm{d}s\right)$$

where I is the conductor current. The expression in brackets is a purely geometrical factor equal to ln D, where D is a 'mean' distance between the observation point P and the conductor. Hence, at point P,

$$A_z = -\frac{\mu_0 I}{2\pi} \ln D.$$

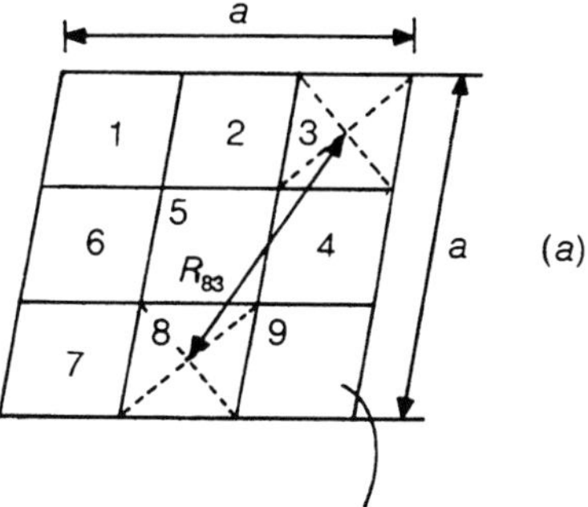

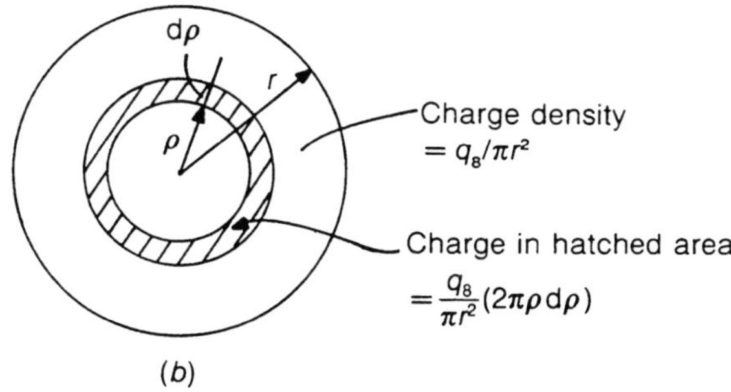

Figure 6.1 (*a*) Division of plate into nine segments ot implement numerical scheme, and (*b*) determination of potential at the centre of a uniformly charged disc.

These four examples show how some of the mathematical tools developed in this chapter can be used. Undoubtedly, these powerful tools can, in the hands of a mathematically competent person, yield useful results. There are, however, severe limitations. Most practical systems are so complex that analytical solutions to Maxwell's equations are either impossible, or of such complexity as to be of limited value. In such cases a numerical solution to these equations is sought. There are, broadly speaking, two classes of numerical techniques which can be used to solve Maxwell's equations.

The first class of numerical methods, known as *integral methods*, are implemented by enforcing certain conditions on boundary surfaces of the problem. An example will show how such methods are generally applied. Let us consider a thin metal conducting sheet placed in free space as shown in Figure 6.1(*a*). An amount of charge Q is deposited on this plate and the problem posed is to find how this charge will distribute on the plate. Depending on one's physical intuition, it may be imagined that the charge distributes uniformly, concentrates near the middle, or near the edges, etc. What is required therefore is a systematic method to determine the magnitude of the charge everywhere on the plate. This is a classic case where the charge distribution is not known and the field cannot be calculated directly using (2.17) and (6.11). The unique feature of this problem is the fact that the surface defined by the conducting sheet is an equipotential. A solution can therefore be found by expressing this physical constraint in mathematical form. In order to do this, resort is made to engineering approximation. It is imagined that the conducting sheet consists of a number of individual segments. Nine such segments are shown in Figure 6.1(*a*). On each segment n it is assumed that there is a charge q_n distributed uniformly. The more segments are chosen the better this approximation is. The choice of the total number is a balance between the required accuracy and resolution, and computational complexity. The essence of the method is in calculating the potential at the centre of each segment due to the charge on each other segment and on itself, and in demanding that all such potentials are equal (since the conducting sheet is an equipotential). The formulae required for this calculation are simple. For example, the potential contribution at segment 8 due to segment 3 is $V_{83} = q_3/4\pi\varepsilon R_{83}$, where R_{83} is the separation between the two segments. The potential contribution at 8 due to q_8 can be found by an approximate formula, by assuming that the segment is circular in shape and of the same area as the square segment, i.e. $\pi r^2 = (a/3)^2$, hence $r = a/(3\sqrt{\pi})$. Hence with reference to Figure 6.1(*b*),

$$V_{88} = \frac{1}{4\pi\varepsilon}\int_{\rho=0}^{r}\frac{2n\rho\,\mathrm{d}\rho}{\rho}\frac{q_8}{\pi r^2} = \frac{q_8}{2\pi\varepsilon r} = \frac{3q_8}{2\sqrt{\pi}\varepsilon a}.$$

Using similar expressions for all nine segments, nine equations can be formulated expressing the condition of constant potential on the conducting sheet. These can then be solved for the unknown charge on each segment. Sophisticated numerical methods based on these ideas have been developed and are known collectively as moment methods.

It should be noted that the formulation of the problem shown in Figure 6.1(*a*) would have been just as easy irrespective of the exact shape of the conducting sheet. This is a very powerful feature of numerical methods permitting a solution to problems which lack the symmetry so important in deriving exact analytical solutions.

The second class of numerical methods are known as *differential methods* and effectively seek a solution to Maxwell's equations in differential form. Unlike the case in integral methods where conditions are imposed on boundary surfaces of

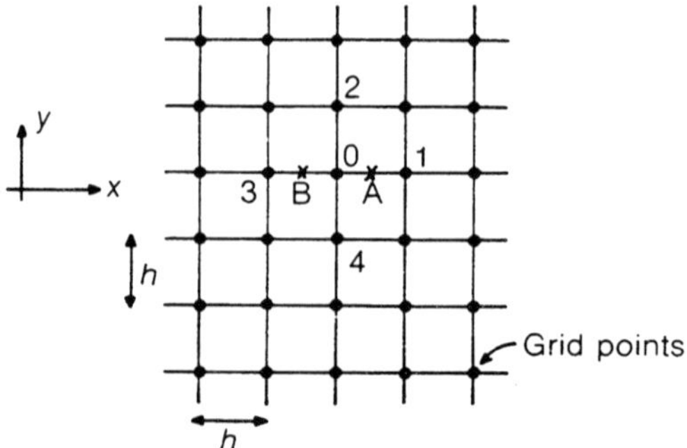

Figure 6.2 Grid used to implement the finite difference scheme.

the problem, in differential methods conditions are imposed on the entire volume of the problem. This normally requires substantial computing resources, but the method is better suited to problems where strong non-uniformities and non-linearities are present. An example of such a method is the finite difference method. Its implementation can be understood by reference to Figure 6.2, where a grid of points is shown in part of the problem space. The main grid points are 0, 1, 2, 3, 4, whilst A, B, are auxiliary points. It is assumed that in this region the potential is required and that it obeys Laplace's equation. The derivative of the potential with respect to x evaluated at point A is

$$\left.\frac{\partial V}{\partial x}\right|_{A} = \frac{V_1 - V_0}{h}$$

where h is the spacing between the main grid points. Similarly,

$$\left.\frac{\partial^2 V}{\partial x^2}\right|_{O} = \frac{(\partial V/\partial x)_A - (\partial V/\partial x)_B}{h} = \frac{1}{h^2}[(V_1 - V_0) + (V_3 - V_0)]$$

and for the y-derivative

$$\left.\frac{\partial^2 V}{\partial y^2}\right|_{O} = \frac{1}{h^2}[(V_2 - V_0) + (V_4 - V_0)].$$

Adding gives

$$\left(\frac{\partial^2 V}{\partial x^2} + \frac{\partial^2 V}{\partial y^2}\right)_{\text{at point O}} = \frac{V_1 - V_0}{h^2} + \frac{V_2 - V_0}{h^2} + \frac{V_3 - V_0}{h^2} + \frac{V_4 - V_0}{h^2}.$$

But from Laplace's equation the left-hand side is equal to zero; hence, after re-arrangement,

$$V_0 = \tfrac{1}{4}V_1 + V_2 + V_3 + V_4. \tag{6.17}$$

In implementing this method the entire problem space is filled with grid points. Fixed potential values are assigned to grid points corresponding to boundaries where the potential is specified (boundary conditions). In the remaining grid points arbitrary potential values are assigned. Equation (6.17) is then applied on all grid points (except those associated with boundary conditions) so that all potential values are recalculated from potential values at their nearest neighbours. This process is repeated until all potentials converge to fixed values. This normally requires three or four such calculations. More sophisticated methods are available to speed up convergence but (6.17) gives a simple first introduction to the technique. Other parameters such as the electric field and the capacitance between electrodes can then be obtained from the potentials.

REMARKS

Maxwell's equations in integral form equations (4.3), (4.4), (2.20) and (6.1) developed from experimental laws) and in differential form (equations (6.4) to (6.8) developed by mathematical manipulation) form a powerful mathematical model of electromagnetic phenomena. The engineer and the mathematician will inevitably use this model in a different way and will have different views of its usefulness. Some examples of using these equations will be given in Part III. In practical problems analytical solutions to these equations are impossible and resort is made to numerical methods. Two such methods were briefly introduced. Other powerful numerical modelling methods are the finite element method based on energy minimization principles, and the transmission-line modelling method based on circuit models for electromagnetic fields.

PROBLEMS

P55 Examine whether the quantity $\mathbf{B} = B_0 e^{at}\hat{z}$, where B_0, a are constants, can represent the magnetic field in a region of space.
Hint: Check whhether **B** as given above can be a solution of Maxwell's equations. Attempt to calculate the electric field using (6.5) and (6.6).
(*Answer*: **B** is not a proper solution of Maxwell's equations.)

P56 Obtain an expression for the electric field in the region between two coaxial electrodes by solving Laplace's equation in cylindrical coordinates. Compare results with those obtained in Example E10.

P57 The electric field in a region of space is given by the formulae

$$E_r(r) = \begin{cases} Ar & \text{for } r < a \\ B/r^2 & \text{for } r > a \end{cases}$$

where A, B are constants. Calculate the charge density everywhere in this region of space.
(*Answer*: $3A\varepsilon$ for $r < a$, zero for $r > a$.)

P58 The density of charged carriers in a junction diode is shown in Figure P58. ρ_+, ρ_- represent the number of particles per m^3, each charge $+e$, $-e$ respectively.

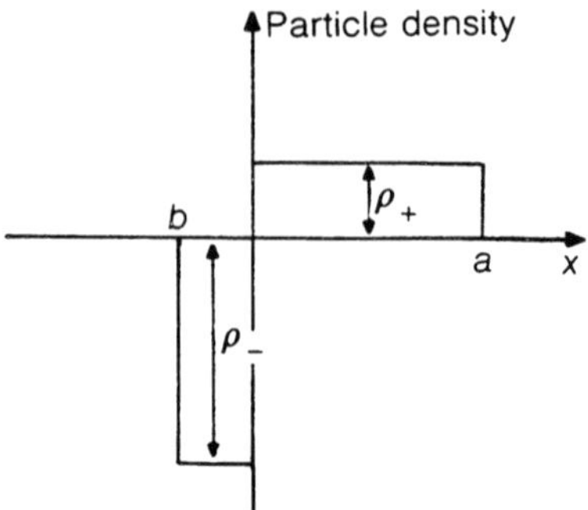

Figure P58

It can be assumed that field components vary along the x-direction only. Solve Poisson's equation to obtain the electric field and the potential difference $V(a) - V(b)$ (the barrier potential). Assume that the total positive and negative charge is the same.
(*Answer*: $V_{ab} = (e/2\varepsilon)(\rho_+ a^2 + \rho_- b^2)$.)

P59 Explore the calculation of the capacitance between two parallel plates by numerical means (integral method).
Hint: select plate separation and cross-sectional area at will, separate each plate into small segments and assign a charge in each segment as explained in the text. Using the formulae in the text obtain expressions for the potential at each segment (there will be contributions from the segment itself, other segments on the same plate and segments on the opposite plate). Impose the condition that the two plates are each an equipotential. Compare results with analytical formulae (which neglect fringing fields). Access to a personal computer is necessary in order to complete this problem.

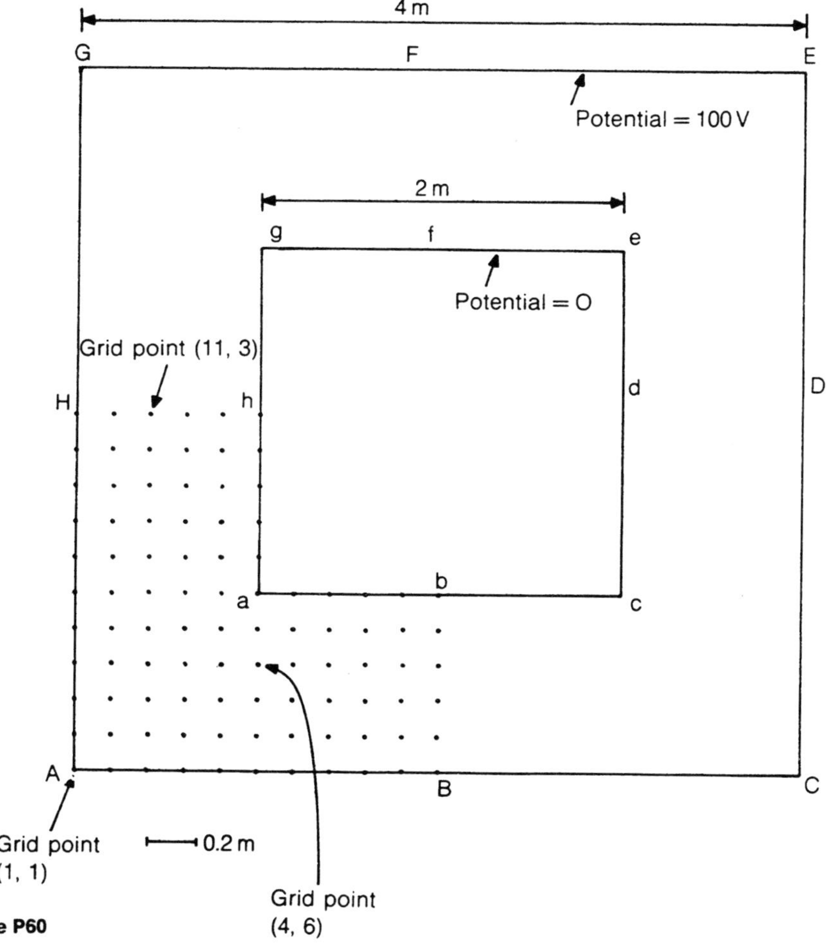

Figure P60

P60 A coaxial cable of square cross-section is shown in Figure P60. The inner conductor abcdefgh is at zero potential, whilst the outer conductor ABCDEFGH is at a potential of 100 V. Use the finite difference routine outlined in the text to obtain the potential distribution between the two conductors and the value of the capacitance per metre length between the two conductors.
Hint: because of symmetry section ABbahHA only need be studied. Write a computer programme where equation (6.17) is repeatedly implemented on all but the boundary nodes of this section. Nodes where boundary conditions are imposed need no further computation. However, nodes along Hh and Bb need special treatment. For example for node (11, 3)

$$V_{(11,3)} = \tfrac{1}{4}(V_{(11,2)} + V_{(11,4)} + V_{(10,3)} + V_{(12,3)})$$

but because of symmetry $V_{(10,3)} = V_{(12,3)}$; hence

$$V_{(11,3)} = \tfrac{1}{4}(V_{(11,2)} + V_{(11,4)} + 2V_{(10,3)}).$$

After a stable potential distribution is established calculate the charge on the inner conductor by using Gauss's law in integral form (the value of the capacitance found by analytical mean is $C \simeq 10.2\varepsilon_0 \mathrm{F\,m^{-1}}$).

PART III

PHENOMENA ASSOCIATED WITH RAPIDLY ACCELERATING CHARGES

In Part II time-varying phenomena were studied on the assumption that the largest physical dimension s of a system and the time period of a signal were such, that the propagation time of electromagnetic disturbances was small compared to the period ($s/v_0 \ll T$, where v_0 is the velocity of propagation). This is tantamount to assuming that electromagnetic interactions between different parts of the system are practically instantaneous. Although v_0 is very high ($v_0 = c = 3 \times 10^8\,\mathrm{m\,s^{-1}}$ in air), it is possible that for high-frequency signals (T very small) and physically large systems (s large), the inequality $s/v_0 \ll T$ does not hold. In this final part we come to examine the implications of the breakdown of this inequality. Let us look at a simple example which demonstrates why the relative magnitude of s/v_0 and T is important. Two conductor segments A, B (spacing s) are shown in Figure 7.1(a). The currents in the two conductors are as shown in Figures 7.1(b, c). The spacing s is chosen so that $s/v_0 = T/2$ where v_0 is the velocity of propagation of disturbances in the medium surrounding the conductors, and T is the period of the waveform representing the current. Let us calculate the force between the two conductors at the instant in time when the current on conductor B has the value I_b. For the conductor spacing chosen, it takes a time interval $T/2$ for the current in conductor A to influence conductor B. Hence the influence experienced by conductor B when $I_B = I_b$ is that caused by the current in conductor A $T/2$ seconds earlier, i.e. I_{a1}. These two currents (I_{a1}, I_b) are of opposite polarity; hence the force is repulsive. If, however, the time it takes for signals to propagate from A to B is neglected,

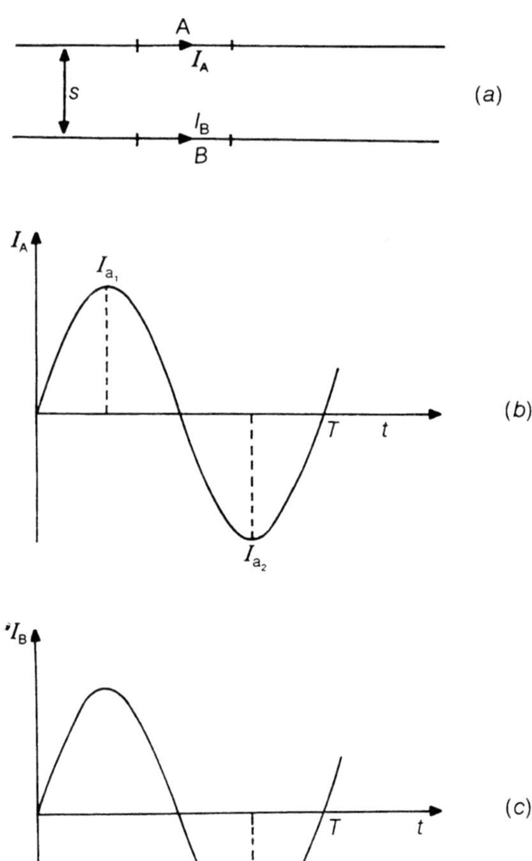

Figure 7.1 Interaction between two adjacent conductor segments A, B.

then the force on B when $I_B = I_b$ is effectively determined by the current on A at the same instant in time i.e. I_{a2}. These two currents (I_b, I_{a2}) are of the same polarity; hence the force turns out to be attractive. Clearly, if $s/v_0 \gtrsim T$ the calculation must be done taking into account finite propagation velocity effects (i.e. using the 'retarded' value of the current in A). If, however, $s/v_0 \ll T$ the propagation effects can be neglected without significant loss in accuracy. In the latter case concepts applicable to static or slowly varying fields (such as capacitance and inductance) can be used to study interactions (section 5.4). It is emphasized again that the important consideration is not the frequency of the signals alone, but the relative magnitude of s/v_0 and $T = 1/f$.

CHAPTER 7

Electromagnetic Waves

The dropping of a pebble into a pond sets up ripples on the surface of the water which appear to propagate away from the source of the disturbance. This type of behaviour is described as wave-like and it is very common in many physical systems (e.g. on a guitar string, electrical transmission line, etc.). In this chapter the possibility of observing wave-like behaviour in the electromagnetic field will be investigated. The reader may question at this point, if he/she has not already done so in connection with the concept of fields, whether there is any physical substance to the notion that waves (like the ripples on the surface of the water) can be associated with electromagnetic fields which propagate even in vacuum. One can think of electromagnetic waves as merely a mathematical model which, since it makes predictions which are in agreement with observations, is acceptable. Making such a limited claim on electromagnetic fields and waves appears to be a pity, but any attempt to assign a deeper physical meaning within accepted notions of space, time and matter presents formidable difficulties. Perhaps the most satisfactory point of view to be adopted by the engineer is to accept that Maxwell's equations and associated concepts are simply a model of reality. As all models, it is incomplete and it is not the same as the real thing. With this philosophical aside let us now define more precisely what are the essential features of waves and electromagnetic waves in particular.

A disturbance propagating along the z-direction with velocity v_0, is shown at times 0 and t in Figure 7.2. A feature of this disturbance is that it propagates undisturbed along z. An observer running along z with velocity v_0 experiences a constant amplitude for the disturbance. The implication of this is that the function f describing the shape of the disturbance depends on $(z - v_0 t)$. Indeed in this case $\mathrm{d}(z - v_0 t)/\mathrm{d}t = (\mathrm{d}z/\mathrm{d}t) - v_0 = v_0 - v_0 = 0$; hence $z - v_0 t = \text{constant}$ and also $f(z - v_0 t) = \text{constant}$ for the observer. A wave is a disturbance which, whilst maintaining its shape, propagates in space. A disturbance such as that shown in Figure 7.3 is not a wave; it is simply an oscillation. It should be noted that a wave cannot be set up if instantaneous propagation of disturbances is assumed. If the source of the disturbance in Figure 7.4 is, say, on oscillator with a sinusoidal output voltage of frequency ω and peak amplitude A then, neglecting any decay in amplitude with distance, the field at a distance z away and at time t is determined

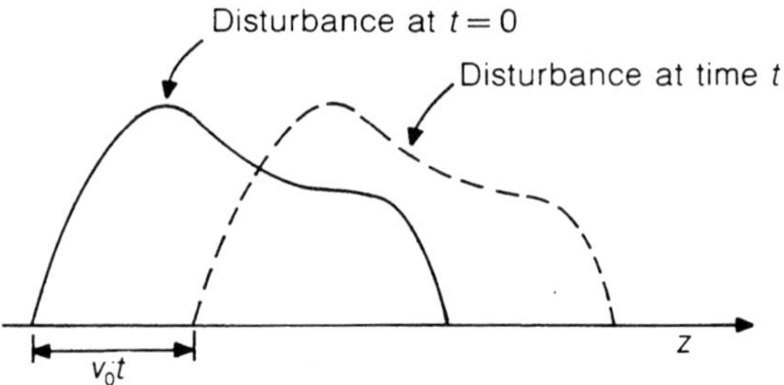

Figure 7.2 The propagation of a wave-line disturbance.

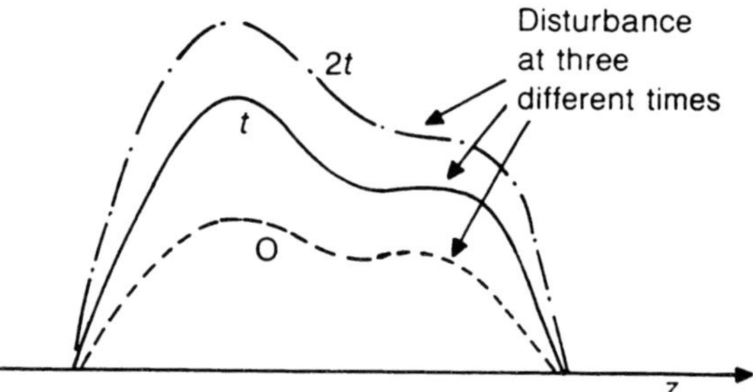

Figure 7.3 The appearance of a disturbance which is not wave-like.

by the source voltage z/v_0 seconds earlier, i.e. $A \sin \omega(t - z/v_0)$. A period of this signal is shown in Figure 7.4. The same signal is plotted Δt seconds later. It can be seen that the waveform at $t + \Delta t$ can be obtained by shifting the waveform at t by a distance $\Delta z = v_0, \Delta t$ i.e. a sinusoidal wave propagating with velocity v_0 has been established. If, however, the velocity of propagation is assumed infinite, no wave-like behaviour is established. It can be imagined that the oscillator is switched on for one period and then it is switched off. The disturbance is then detached from the source and propagates in space towards its destination. Whilst the disturbance is airborne so to speak, en route to its destination, it has no connection with the source and it transports energy. It is this entity which perhaps is regarded as the electromagnetic field.

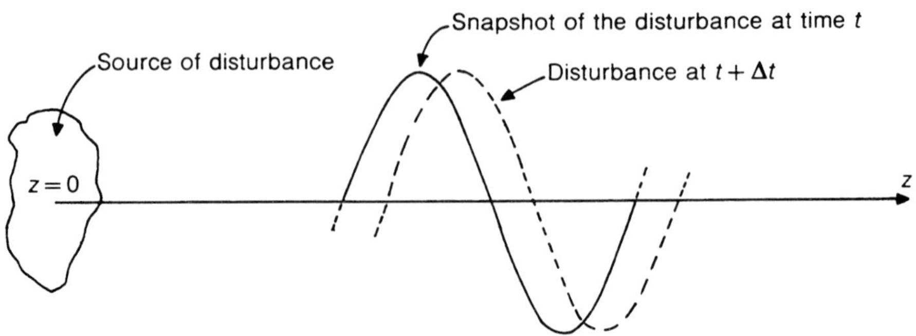

Figure 7.4 Launching of a wave-like disturbance.

A solution to Maxwell's equations is sought which shows a wave-like behaviour as described above. The simplest form of such a solution is when the electric field has only a y-component, $\mathbf{E} = (0, E_y, 0)$, the magnetic field only a z-component $H = (0, 0, H_z)$, and the only variation allowed is along the x-direction (one-dimensional plane wave). This is a natural choice for a wave propagating along the x-direction, since Poynting's vector, which represents power flow, is in this case $\mathbf{E} \times \mathbf{H} = E_y H_z \hat{\mathbf{x}}$. Let us see whether such a wave solution can be obtained from Maxwell's equations. Using these equations in cartesian form (see Appendix B)

$$\nabla \times \mathbf{E} = -\frac{\mu \partial \mathbf{H}}{\partial t} \qquad \text{reduces to} \qquad \frac{\partial E_y}{\partial x} = -\mu \frac{\partial H_z}{\partial t} \tag{7.1}$$

and after neglecting the conduction current

$$\nabla \times \bar{\mathbf{H}} = \bar{\mathbf{J}}_f + \varepsilon \frac{\partial \mathbf{E}}{\partial t} \qquad \text{reduces to} \qquad -\frac{\partial H_z}{\partial x} = \varepsilon \frac{\partial E_y}{\partial t}. \tag{7.2}$$

Differentiating (7.1) with respect to x and (7.2) with respect to t gives

$$\frac{\partial^2 E_y}{\partial x^2} = -\mu \frac{\partial^2 H_z}{\partial x \partial t} \qquad \text{and} \qquad -\frac{\partial^2 H_z}{\partial x \partial t} = \varepsilon \frac{\partial^2 E_y}{\partial t^2}.$$

Eliminating terms in H_z gives

$$\frac{\partial^2 E_y}{\partial x^2} = \mu\varepsilon \frac{\partial^2 E_y}{\partial t^2}. \tag{7.3}$$

An identical equation is obtained for H_z. Equation (7.3) is known as the wave equation and it is easy to show that it admits solutions of the form $E_y(x \pm v_0 t)$. Indeed for $E_y(x - v_0 t)$ $\partial E_y/\partial x = E_y'$ (where the prime indicates differentiation with respect to $(x - v_0 t)$). Similarly $\partial^2 E_y/\partial x^2 = E_y''$, $\partial E_y/\partial t = -v_0 E_y'$ and $\partial^2 E_y/\partial t^2 = v_0^2 E_y''$. Substituting into the wave equation gives $E_y'' = \mu\varepsilon v_0^2 E_y''$.

This expression indicates that wave solutions are possible with a wave propagation velocity $v_0 = 1/\sqrt{\mu\varepsilon}$. For propagation in air $v_0 = c = 3 \times 10^8\,\mathrm{m\,s^{-1}}$ as already discussed in section 5.4. Solutions of the type $E_y(x + v_0 t)$ represent waves propagating along the negative x-direction. The most interesting practical case is of sinusoidal waves propagating along x. In this case

$$\begin{aligned} E_y(x, t) &= E_0 \sin[\omega(t - x/v_0)] \\ H_z(x, t) &= H_0 \sin[\omega(t - x/v_0)]. \end{aligned} \tag{7.4}$$

These two waveforms are shown at an instant of time in Figure 7.5. The distance λ between two points along x with the same magnitude and phase of E_y is called the wavelength. Obviously, $\omega\lambda/v_0 = 2\pi$ or, since $\omega = 2\pi f$,

$$\lambda = v_0/f. \tag{7.5}$$

Differentiating E_y and H_z in (7.4) gives

$$\frac{\partial E_y}{\partial x} = -\frac{\omega}{v_0} E_0 \cos[\omega(t - x/v_0)] \qquad \frac{\partial H_z}{\partial t} = \omega H_0 \cos[\omega(t - x/v_0)]$$

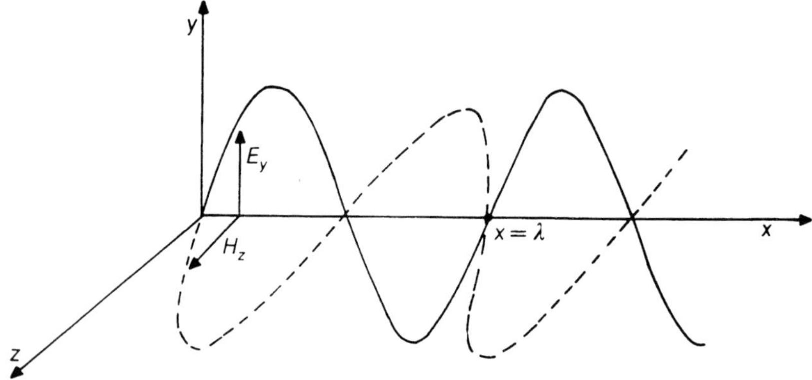

Figure 7.5 The electric and magnetic field of an electromagnetic plane wave.

and substituting in (7.1) results in

$$\frac{E_0}{H_0} = \mu v_0 = \mu \frac{1}{\sqrt{\mu\varepsilon}} = \sqrt{\frac{\mu}{\varepsilon}}.$$

The quantity $Z_0 = \sqrt{\mu/\varepsilon}$ is called the intrinsic impedance of the medium and for air it is equal to 377 Ω. The peak values of electric and magnetic field for this plane wave are related by Z_0:

$$\frac{E_0}{H_0} = Z_0 = \sqrt{\frac{\mu}{\varepsilon}}. \tag{7.6}$$

The power flow associated with this wave is obtained from Poynting's vector, $E_y H_z = E_0^2/2Z_0$ (in W m^{-2}) where the factor of two appears since the rms value of E_y and H_z must be used. Equation (7.5) is an example of what is called a 'dispersion relation'. The wavenumber $k = 2\pi/\lambda$ is also defined, in which case $E_y = E_0 \sin(\omega t - kx)$. A point of constant phase propagates with a velocity V_{ph} (the phase velocity) obtained by demanding that $\omega t - kx =$ constant. Differentiating this expression with respect to t gives $\omega - k\,\mathrm{d}x/\mathrm{d}t = 0$ and hence

$$V_{ph} = \frac{\mathrm{d}x}{\mathrm{d}t} = \frac{\omega}{k}. \tag{7.7}$$

The dispersion relation simply states that, in his case, the phase velocity is constant independent of frequency and equal to $\sqrt{\mu/\varepsilon}$. A pulse-like disturbance propagating in this medium can be represented by a combination of sinusoidal waves of different frequencies (analysis into a Fourier series). Since according to (7.5) all frequency components propagate at the same velocity, they retain their relative phase at any distance away from the launch point. The pulse-like disturbance thus suffers no distortion—there is no dispersion of the pulse. There are cases of propagation in different media where this is not the case (e.g. in the ionosphere).

Another aspect of waves is their polarization. In the derivation of (7.1) and (7.2) only E_y and H_z field components are envisaged. This is an example of a linearly polarized wave. It is possible to obtain waves where the electric field vector consists of components E_y, E_z out of phase to each other so that as the wave propagates $|E|$ remains constant but the field vector rotates about the direction of propagation (elliptically polarized wave).

Electromagnetic waves do not only propagate in free space. Propagation in transmission lines, and inside metallic pipe-like structures (waveguides) is possible. In any case, waves encounter obstacles and as discussed in section 5.4 they suffer reflections. Any termination which presents an impedance different from the intrinsic impedance Z_0 of the surrounding medium will result in reflections. The outcome of these reflections is wave patterns which lack the simplicity expressed by equations (7.4). Nevertheless these waves can be studied, but inevitably they result in greater mathematical complexity. An interesting aspect of this problem is brought about by the fact that waves have definite spatial features characterized by their wavelength λ. An analogy with vibrations on a guitar string will help to illustrate this point. Figure 7.6 shows three of the possible modes of vibration of the string. They are determined by the requirement that the displacement at the fixing points is zero. Similarly a wave established inside a metal box (in a microwave oven, for example) is constrained to have its electric field component parallel to the conducting surfaces approaches zero near such surfaces. Any wave set up inside the box is restricted by these constraints and, as a result, there are certain preferred frequencies (called 'modes') at which organized field patterns can be established. A detailed study of these matters falls within the scope of waveguide theory.

Finally, it remains to examine briefly the state of affairs as regards wave propagation inside a material which is lossy. Returning to equations (7.1) and (7.2) but this time retaining the conduction current gives

$$\frac{\partial E_y}{\partial x} = -\mu \frac{\partial H_z}{\partial t}$$

$$-\frac{\partial H_z}{\partial x} = j_y + \varepsilon \frac{\partial E_y}{\partial t} = \sigma E_y + \varepsilon \frac{\partial E_y}{\partial t}.$$

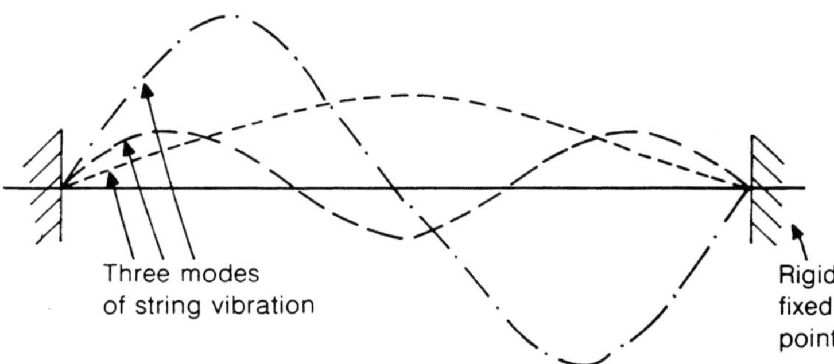

Figure 7.6 Modes in a vibrating string.

Differentiating the first equation with respect to x, the second with respect to t, and eliminating terms in H_z results in

$$\frac{\partial^2 E_y}{\partial x^2} = \mu\varepsilon\frac{\partial^2 E_y}{\partial t^2} + \mu\sigma\frac{\partial E_y}{\partial t}.$$

If losses are neglected ($\sigma = 0$) the wave equation (7.3) is obtained. If, on the other hand, the first term on the right-hand side is negligible the resulting equation is known as the diffusion equation and it predicts behaviour not unlike that found in section 5.2 for the penetration of fields inside a conductor. Whether wave-like or diffusion-like behaviour dominates depends on the relative magnitude of these two terms. Assuming for E_y a sinusoidal variation with frequency gives

$$\left|\mu\varepsilon\frac{\partial^2 E_y}{\partial t^2}\right| \sim \mu\varepsilon\omega^2 \qquad \left|\mu\sigma\frac{\partial E_y}{\partial t}\right| \sim \mu\sigma\omega.$$

Clearly, wave-like effects dominate when $\omega\varepsilon \gg \sigma$ (i.e. at high frequencies and for poor conductors). At low frequencies and for good conductors the diffusion term is dominant.

EXAMPLE E34

For the coaxial transmission line shown in Figure E34 determine whether wave-like propagation is possible with a component of the electric field in the r-direction only, given by

$$E_r = \frac{E_0}{r}\sin\left[\omega(t - z/v_0)\right].$$

Calculate all other parameters of the field and the conductor current and charge.

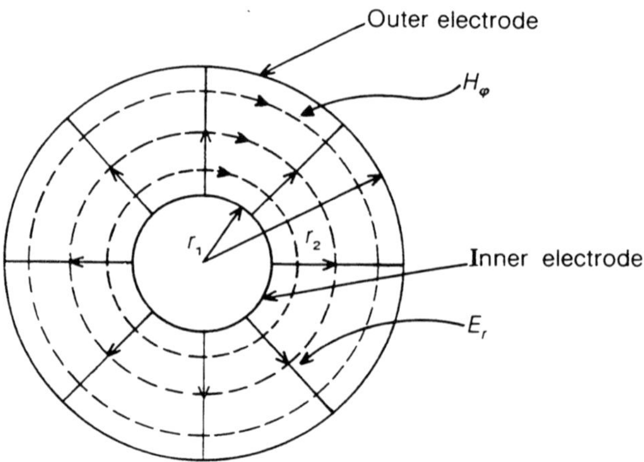

Figure E34

Solution

This is a problem with cylindrical symmetry, hence variations in the azimuthal φ-direction are zero. The wave equation in a more general form is found after some algebra, or from analogy to (7.3), to be

$$\nabla^2 \mathbf{E} = \mu\varepsilon \frac{\partial^2 \mathbf{E}}{\partial t^2}. \tag{7.8}$$

In cylindrical coordinates, for E_r component only, and no variation in φ, (7.3) reduces to

$$\nabla^2 \mathbf{E} = \nabla^2 E_r \hat{\mathbf{r}} = \hat{\mathbf{r}}\left[\frac{1}{r}\frac{\partial}{\partial r}\left(r\frac{\partial E_r}{\partial r}\right) + \frac{\partial^2 E_r}{\partial z^2} - \frac{E_r}{r^2}\right] \qquad \text{(see appendix B)}$$

$$= \hat{\mathbf{r}}\left[\frac{1}{r}\frac{\partial E_r}{\partial r} + \frac{\partial^2 E_r}{\partial r^2} + \frac{\partial^2 E_r}{\partial z^2} - \frac{E_r}{z^2}\right]$$

but

$$\frac{\partial E_r}{\partial r} = -\frac{E_0}{r^2}\sin[\omega(t - z/v_0)]$$

$$\frac{\partial^2 E_r}{\partial r^2} = \frac{2rE_0}{r^4}\sin[\omega(t - z/v_0)]$$

$$\frac{\partial^2 E_r}{\partial z^2} = -\left(\frac{\omega}{v_0}\right)^2 \frac{E_0}{r}\sin[\omega(t - z/v_0)].$$

Hence

$$\nabla^2 \mathbf{E} = -\left(\frac{\omega}{v_0}\right)^2 \frac{E_0}{r}\sin[\omega(t - z/v_0)]\hat{\mathbf{r}}$$

and similarly

$$\frac{\partial^2 \mathbf{E}}{\partial t^2} = -\omega^2 \frac{E_0}{r}\sin[\omega t - z/v_0)]\hat{\mathbf{r}}.$$

Clearly the wave equation is satisfied provided $v_0 = 1/\sqrt{\mu\varepsilon}$. The magnetic field associated with this wave can be obtained from $\nabla \times \mathbf{E} = -\partial \mathbf{B}/\partial t$, which for the circumstances of this problem gives

$$\nabla \times \mathbf{E} = -\frac{\partial E_r}{\partial \mathbf{z}}\hat{\boldsymbol{\varphi}}$$

Hence, there is only a φ-component of the magnetic field such that

$$-\mu\frac{\partial H_\varphi}{\partial t} = -\frac{\partial E_r}{\partial z}$$

and therefore

$$H_\varphi = \frac{E_0}{\mu r v_0}\sin[\omega(t - z/v_0)]$$

as can directly be confirmed by substitution.

The power flux can also be found from Poynting's vector, $\mathbf{P} = \mathbf{E} \times \mathbf{H} = E_r H_\varphi \hat{\mathbf{z}}$ which indicates power flow along the z-direction. Substituting field values gives

$$P = \frac{E_0^2}{\mu r^2 v_0} \sin^2[\omega(t - z/v_0)].$$

The average power flow is obtained by averaging this expression in time, which simply leads to $P_{\text{ave}} = (E_0^2/2\mu r^2 v_0)$ (in W m^{-2}). The total power flow over the entire space between the conductors where the wave is established is obtained by integration:

$$P_{\text{total}} = \frac{E_0^2}{2\mu v_0} \int_{r=r_1}^{r_2} \frac{1}{r^2} 2\pi r dr = \frac{\pi E_0^2}{\mu v_0} \ln \frac{r_2}{r_1} \qquad \text{(W)}.$$

It is also interesting to calculate the current and charge distributions on the inner conductor. The current can be obtained by applying Ampere's law on a circular curve of radius r just enclosing the inner conductor ($r \simeq r_1$). The current is then found to be $I = 2\pi r_1 H_\varphi$. Similarly, the charge density on the conductor can be obtained from the appropriate boundary condition $\sigma_f = \varepsilon E_r$. Both conductor and charge distributions show wave-like behaviour corresponding to the field component associated with them. Similar conclusions can be drawn for the current and charge distributions on the outer conductor.

In section 5.4 it was shown that a voltage disturbance V_0 travelling down a line has an associated current disturbance $I_0 = V_0/Z$, where Z is the characteristic impedance of the line. Hence, since the current on the inner conductor has just been calculated, it is possible to obtain the power flow down the line in terms of circuit quantities, $P = V_0 I_0 = I_0^2 Z$, where I_0 is the rms value of the conductor current $I_0 = 2\pi r_1 (E_0/\sqrt{2}\mu r_1 v_0)$. The characteristic impedance of the line is $Z = \sqrt{L_e/C_e}$ where C_e and L_e are the capacitance and inductance per unit length calculated in Examples E10 and E20 respectively. Substituting these values gives

$$Z = \left[\frac{\mu}{2\pi}\left(\ln\frac{r_2}{r_1}\right)^2 \frac{1}{2\pi\varepsilon}\right]^{1/2} = \frac{1}{2\pi} \ln\frac{r_2}{r_1} \sqrt{\frac{\mu}{\varepsilon}}.$$

(Note that $Z_0 = \sqrt{\mu/\varepsilon}$ is the impedance of the medium relating E and H, whilst Z is the impedance of the line relating V and I.)

Hence

$$P = \left(2\pi \frac{E_0}{\sqrt{2}\mu v_0}\right)^2 \frac{1}{2\pi} \ln\frac{r_2}{r_1} \sqrt{\frac{\mu}{\varepsilon}} = \frac{\pi E_0^2 \ln(r_2/r_1)}{\mu v_0}.$$

This formula is identical to that obtained from the field description of the problem using Poynting's vector. The two descriptions (circuit and field) are equivalent. Which description is used depends on the nature of the problem studied.

REMARKS

The rudiments of wave propagation were presented in this chapter. It was shown how wave-like solutions to Maxwell's equations can be obtained and how some of the fundamental wave parameters can be determined. Wave propagation covers a vast range of frequencies and practical applications (dc to light). However, many of the essential aspects of waves do not depend on frequency and the ideas explored in this chapter find a very wide application. It has been stated that truly high-frequency phenomena, which require a different kind of thinking to low-frequency phenomena, occur when the propagation time across the system studied is comparable to or larger than the period of the electrical disturbance. This criterion can also be expressed by comparing the largest physical dimension s of the system with the wavelength λ of the disturbance. If $s \ll \lambda$ interactions are practically instantaneous and low-frequency (quasistatic) behaviour is evident. When, however, $s \gtrsim \lambda$ high-frequency type phenomena are observed. The means by which waves are launched and energy is radiated away from a source are examined in the next chapter.

PROBLEMS

P61 A uniform plane wave propagates in a lossless medium ($\varepsilon_r = 2.5$, $\mu_r = 1$). The peak amplitude of the electric field is $75\ \mathrm{V\,m^{-1}}$ and the frequency is 95 MHz. Calculate:

(*a*) the velocity of propagation in this medium;
(*b*) the intrinsic impedance of the medium;
(*c*) the wavelength;
(*d*) the peak amplitude of the magnetic field;
(*e*) the power flux.
(*Answer:* (*a*) $1.89 \times 10^8\ \mathrm{m\,s^{-1}}$, (*b*) $238\ \Omega$, (*c*) 1.99 m, (*d*) $0.315\ \mathrm{A\,m^{-1}}$, (*e*) $11.8\ \mathrm{W\,m^{-2}}$).

P62 The magnetic field component of a uniform plane wave is given by the formula $H_z = H_0 \sin(\omega t - kx)$, and is plotted at time $t = 0$ in Figure P62. Obtain a formula for the corresponding electric field component and plot its spatial variation at $t = 0$.

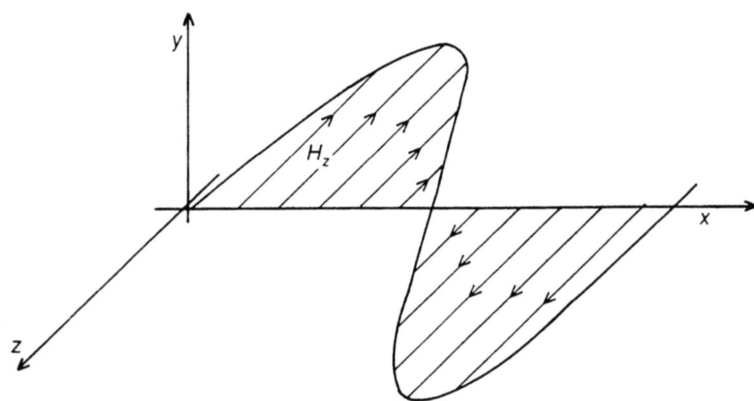

Figure P62

P63 A plane wave is polarized with the E vector in the y-direction and is propagating along the z-direction. The frequency is 95 MHz and the velocity of propagation is 2×10^8 m s^{-1}. E_y reaches its positive peak value of 2 V m^{-1} at $t = 0$, $z = 5$ m. Obtain a formula for E_y at $z = 8$ m.
(*Answer:* $E_y(8, t) = 2\sin(5.97 \times 10^8\, t + \pi/2 - 8.95)$.)

P64 (*a*) Show that the equation $(\partial^2 \bar{E}_y/\partial x^2) = \mu\varepsilon(\partial^2 \bar{E}_y/\partial t^2) + \mu\sigma(\partial \bar{E}_y/\partial t)$ admits solutions of the type $\bar{E}_y \sim e^{j(\omega t - kz)}$ where $k = \omega\sqrt{\mu\varepsilon}\sqrt{1 - j(\sigma/\omega\varepsilon)}$.
(*b*) For low-loss materials ($\sigma/\omega\varepsilon \ll 1$) show that the wave decays with distance as $\exp(-\frac{1}{2}\sigma\sqrt{\mu/\varepsilon}\, z)$.
(*c*) For very lossy materials ($\sigma/\omega\varepsilon \gg 1$) show that the wave decays as $\exp(-\sqrt{\omega\mu\sigma/2}\, z)$.
(*d*) Suggest a frequency suitable for communication with submarines to a depth of 75 m, the sea parameters being $\varepsilon_r = 80$, $\mu_r = 1$, $\sigma = 4.3$ S m^{-1}.
(*Answer:* (*d*) 10 Hz.)

P65 Show that the electric field components $E_y = E_0 \sin(\omega t - kx)$ and $E_z = E_0 \cos(\omega t - kx)$ represent a wave travelling along the x-direction with peak amplitude $\sqrt{2}E_0$ and a total electric field vector making an angle with the z-axis equal to $\omega t - kx$ (circularly polarized wave).

CHAPTER 8

Radiating Systems

The generation of electromagnetic waves is achieved by exciting suitable structures, called antennas, using signals from high-frequency sources. In the source region, which typically consists of an antenna fed from a signal generator via a transmission line, charge and current distributions are established. The scalar and vector potentials, $V(\mathbf{r}, t)$ and $\mathbf{A}(\mathbf{r}, t)$ respectively, at a point a distance $\mathbf{r}$ away from the source (Figure 8.1) are determined by the source current and charge distributions at an earlier time $t - r/v_0$ where v_0 is the velocity of propagation of the disturbance in the medium surrounding the source region. Hence, formulae (2.17) and (6.11) must be modified to account for the propagation delay

$$V = \frac{1}{4\pi\varepsilon}\int \frac{\rho_v(t - r/v_0)}{r}\mathrm{d}v \tag{8.1}$$

$$\mathbf{A} = \frac{\mu}{4\pi}\int \frac{\mathbf{j}_f(t - r/v_0)}{r}\mathrm{d}v. \tag{8.2}$$

These formulae give the 'retarded' scalar and vector potentials and must be used in cases where propagation delays cannot be ignored. As long as ρ_v and $\mathbf{j}_f$ are known, V and $\mathbf{A}$ and hence $\mathbf{E}$ and $\mathbf{B}$ can be calculated. The nature of the propagating wave can then be assessed and its parameters calculated (e.g. the radiated power can be found from Poynting's vector). Clearly not all circuits are efficient radiators. In the remainder of this chapter the features desirable in an efficient radiator are examined and examples of calculating the radiated fields are given.

Let us consider the circuit shown in Figure 8.2(*a*), where a high-frequency signal source feeds a small capacitor via a two-wire transmission line. The reader will have no difficulty in accepting that a current will flow in this circuit. If the size of the plates is reduced and the spacing is increased inevitably the capacitance will decrease and hence the current will also decrease. By opening up the two lines and reducing the size of the plates the configuration shown in Figure 8.2(*b*) is eventually reached. Electrons in the two wire segments will still be set in motion and current will flow at high frequencies. Let us also arrange so that the length l is equal to half the wavelength corresponding to the frequency of the signal source. The current must be zero at the extremities of the wire, so that at the moment the source current is at its

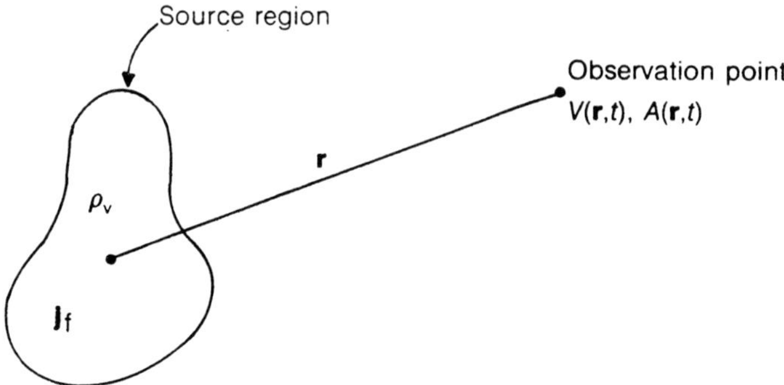

Figure 8.1 Determination of scalar and vector potential due to a general charge/current distribution.

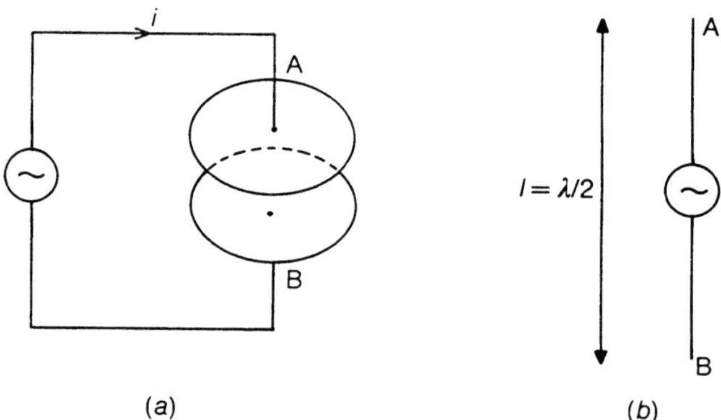

Figure 8.2 Development of a radiating structure.

positive peak the current distribution along the wire will be as shown in Figure 8.3(*a*). (In this figure the length of the arrows indicates the magnitude of the current.) It should be noted that in cases such as this, where the physical size of the circuit in Figure 8.2(*b*) is comparable to the wavelength, the current is not constant along the wire. This is in marked contrast to the low-frequency case ($l \ll \lambda$) where a constant current at any cross-section of the wire will be expected. For the situation depicted in Figure 8.3(*a*) (I positive maximum) there are no charge imbalances anywhere in the wire and hence there is zero net charge (in analogy with the circuit in Figure 8.2(*a*) where when $i = I_{max}$ then $v = 0$, $q = 0$). A quarter of a period later the current will be zero and the net charge along the wire will be as shown in Figure 8.3(*b*). The reader may find it difficult to accept this charge distribution. After all, charges of opposite polarity attract each other and hence one would expect charges to concentrate near the middle point. This difficulty is resolved if one considers that the interaction experienced by the positive charges at the top of the wire is that due to charges at the lower end of the wire half a period earlier (since it takes $T/2$ seconds for a signal to traverse the half wavelength distance between the upper and lower ends of the wire). These charges are positive, so the force acting on the top charge is repulsive. This is another example of the significance of the finite

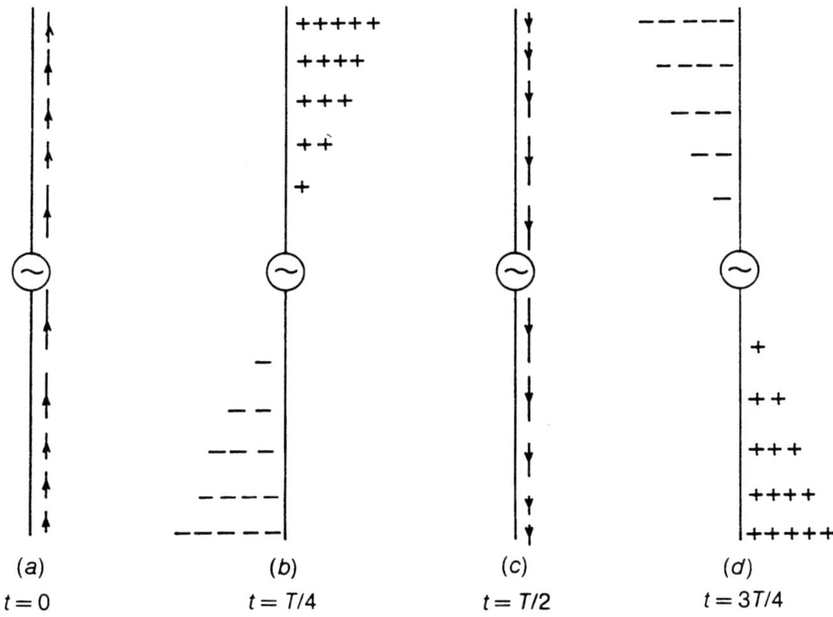

Figure 8.3 Current and charge distribution on a dipole at different times.

propagation velocity in this kind of situation. Figures 8.3(*c*) and (*d*) show the next two stages in the process. This device is a very common type of antenna known as the 'half-wavelength dipole'. To calculate the radiation field pattern due to this antenna, the scalar and vector potentials are obtained from (8.1) and (8.2) and the field components are calculated in the normal way (equations (6.12) and (6.13)). At large distances from this dipole ($r \gg \lambda$), in what is called the far-field region, the following field components are found:

$$|E_\vartheta| \simeq \sqrt{\frac{\mu}{\varepsilon}}\frac{I_0}{2\pi r}\left|\frac{\cos[(\pi/2)\cos\vartheta]}{\sin\vartheta}\right| \qquad (\mathrm{V\,m^{-1}}) \qquad (8.3)$$

and

$$H_\varphi \simeq \frac{|E_\vartheta|}{\sqrt{\mu/\varepsilon}} \qquad \text{for} \quad r \gg \lambda.$$

The field pattern is shown in Figure 8.4. The radiated power is stronger in certain directions. This feature is known as the 'directivity' of the antenna and much effort is put into designing antennas with the desired directivity. Another important parameter in antenna design is the radiation resistance, defined as the ratio of the total radiated power over the square of the rms current supplied by the source. For a half-wavelength dipole $R_r = 73\,\Omega$. The transmission line feeding this antenna must have a characteristic impedance which is matched by R_r. In general, the effective impedance seen by the feed line is the radiation resistance plus a reactance which, depending on the circumstances, is capacitive or inductive in nature. The proximity of conducting bodies (e.g. earth) or other dipoles affects the current distribution and

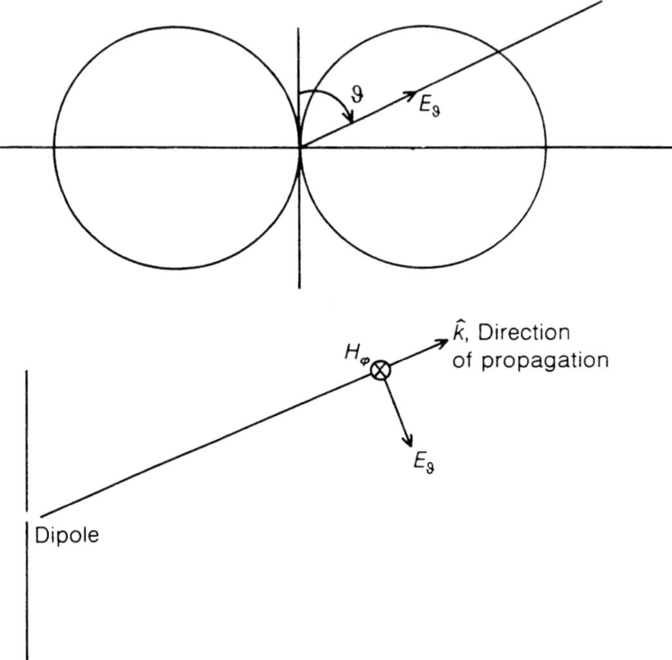

Figure 8.4 E_θ pattern for a half-wavelength dipole.

hence modifies all the antenna characteristics. All these effects can to some extent be quantified, normally with the use of sophisticated numerical techniques.

An interesting feature of (8.3) is that both E_ϑ and H_φ decay with distance as $1/r$. This is typical of radiation behaviour, as opposed to the decay of electrostatic fields which scale as $1/r^2$.

An efficient radiating structure has dimensions of the order of a wavelength. If spacings are much smaller than a wavelength, opposite currents from different parts of the circuit produce effects which cancel each other out. Hence, such elements must be kept well apart. There are, in addition to dipole antennas, many other types, e.g. loop antennas, horn antennas, and arrays of elementary antennas arranged in such a way as to achieve better directivity.

EXAMPLE E35

Determine the radiation field on the $\vartheta = \pi/2$ plane of the very short dipole shown in Figure E35(a).

Solution

An immediate issue to be addressed is what is meant by the term 'very short dipole' and what current distribution is appropriate for such a structure. Clearly the length

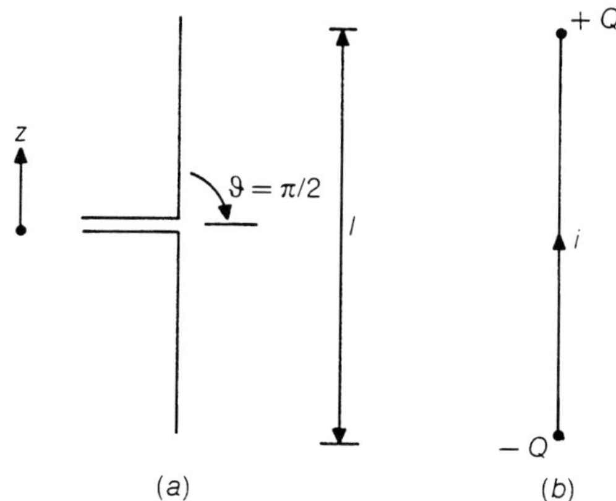

Figure E35

l of the dipole must be much smaller when compared to the wavelength. A value $l < \lambda/50$ is acceptable. Under such circumstances the current along the dipole can be regarded as constant $i(t) = I\cos(\omega t)$, as shown in Figure E35(*b*). Continuity requires that the charge Q is such that $dQ/dt = i$, hence $Q = (I/\omega)\sin(\omega t)$. The strategy of the calculation is to obtain the scalar potential (due to $\pm Q$) and the vector potential (due to I), and then to calculate **E**, **B**. By restricting the calculation to the $\vartheta = \pi/2$ plane, mathematical complexity is reduced since $V(\vartheta = \pi/2) = 0$ (equal and opposite contributions from $\pm Q$). The vector potential need only be calculated. Using (8.2)

$$A_z(\vartheta = \pi/2) = \frac{\mu I \cos[\omega(t - r/v_0)]}{4\pi r} l.$$

Since the source current is in the z-direction only, all other vector potential components are zero. Hence,

$$E_z(\vartheta = \pi/2) = -\frac{\partial A_z}{\partial t} = \frac{\mu I l}{4\pi r}\omega \sin[\omega(t - r/v_0)].$$

All other electric field components are zero.

The magnetic field is similarly obtained:

$$\mathbf{B} = \nabla \times \mathbf{A} = -\frac{\partial A_z}{\partial r}\hat{\boldsymbol{\varphi}}$$

and hence

$$H_\varphi = \frac{Il}{4\pi}\left(-\frac{\omega \sin[\omega(t - r/v_0)]}{v_0 \qquad r} - \frac{1}{r^2}\cos[\omega(t - r/v_0)]\right).$$

At long distances away from the dipole, the second term in this expression becomes very small (since it varies as $1/r^2$). Far from the dipole only the terms dependent on $1/r$ remain. The power flow at the $\vartheta = \pi/2$ plane can be calculated from Poynting's vector $P = E_z H_\varphi$. When this is done, the first term in H_φ results in an expression proportional to $\sin^2(\cdots)$, and when averaged indicates a net power flow away from the dipole. However, the second term contributes an expression proportional to sin(···) cos(···) which always averages to zero. It does not therefore represent net power low—it is effectively a reactive term. It is some distance away (in the 'far-field' region of the antenna) where the field pattern shows purely radiation type behaviour. Close to the radiator (in the 'near-field' region) radiation type behaviour is mixed up with reactive behaviour. Near the radiating structure the field retains a memory of its manner of production. If the radiating structure generates significant potential differences (as in a dipole) electrostatic type terms are evident, whilst if current flow is substantial (as in a loop antenna) magnetostatic type terms predominate. For the example given here, the radiation and reactive terms become equal in magnitude at a distance obtained from $\omega/v_0 r = 1/r^2$, i.e. $r = \lambda/2\pi$. The region around the antenna can thus be classified as follows:

(i) $0 < r < \lambda/2\pi$. Reactive near-field region. Reactive terms predominate. Full radiation pattern (e.g. directivity) not yet established.

(ii) $\lambda/2\pi < r <$ a few wavelengths. Radiation near-field region. Pronounced radiation term present. Evidence of the directional properties of the antenna begin to appear.

(iii) $r >$ a few wavelengths. Radiation far field region. Radiation terms dominant. Directional properties of the antenna fully established. (For large radiation structures of a typical dimension D, this region is deemed to start when $r > 2D^2/\lambda$.)

After some algebra, it is possible to derive expressions for the field components generated from a very short dipole, valid anywhere in the space surrounding the dipole. These are:

$$\bar{E}_r = \frac{p\cos\vartheta}{2\pi\varepsilon}\left[\frac{j\beta}{r^2} + \frac{1}{r^3}\right]$$

$$\bar{E}_\varphi = 0$$

$$\bar{E}_\vartheta = \frac{p\sin\vartheta}{4\pi}\left[\frac{-\omega^2\mu}{r} + \frac{j\beta}{\varepsilon r^2} + \frac{1}{\varepsilon r^3}\right]$$

$$\bar{B}_\varphi = \frac{\mu I l}{4\pi}\sin\vartheta\left[\frac{j\beta}{r} + \frac{1}{r^2}\right]$$

where

$$p = -(j/\omega)Il \quad \text{and} \quad \beta = 2\pi/\lambda.$$

These equations give effectively the phasors representing the field components. Real measurable quantities are obtained by multiplying these expressions by $\exp[j(\omega t - 2\pi r/\lambda)]$ and calculating the real part of the result. It is evident that there are present in the field, radiation terms $(1/r)$ and reactive terms $(1/r^2, 1/r^3)$. As r increases, the far-field region is reached and the only field components present are

$$\bar{E}_\vartheta = -\frac{p\sin\vartheta}{4\pi r}\omega^2\mu$$

$$\bar{H}_\varphi = \frac{\bar{B}_\varphi}{\mu} = \frac{p\sin\vartheta}{4\pi r}\omega^2\sqrt{\mu\varepsilon}.$$

It is clear from these expressions that $E_\vartheta = H_\varphi\sqrt{\mu/\varepsilon}$ and that a fixed field pattern has been established. Power flow is in the radial direction and can be calculated from Poynting's vector:

$$P = E_{\vartheta} H_{\varphi} = -\left(\frac{p \sin \vartheta}{4\pi r}\right)^2 \omega^4 \mu\varepsilon \sqrt{\frac{\mu}{\varepsilon}} \frac{1}{2}$$

$$= \frac{\omega^2 \mu l^2 I^2}{32\pi^2 r^2 v_0} \sin^2 \vartheta \qquad (\mathrm{W\,m^{-2}})$$

where I is the peak current and the factor 1/2 appearing in the second substitution is there to convert to rms values. This power flux pattern is shown in Figure 8.5. The total radiated power can be calculated by integrating the power flux over a spherical area surrounding the dipole. Using the surface element in Figure 8.6

$$P_{\text{total}} = \int_{\vartheta=0}^{\pi} \int_{\varphi=0}^{2\pi} P r^2 \sin \vartheta \mathrm{d}\varphi \, \mathrm{d}\vartheta = \frac{\omega^2 \mu I^2 l^2}{12\pi v_0} \qquad (\mathrm{W}).$$

As far as the circuit feeding the antenna is concerned, the power radiated appears

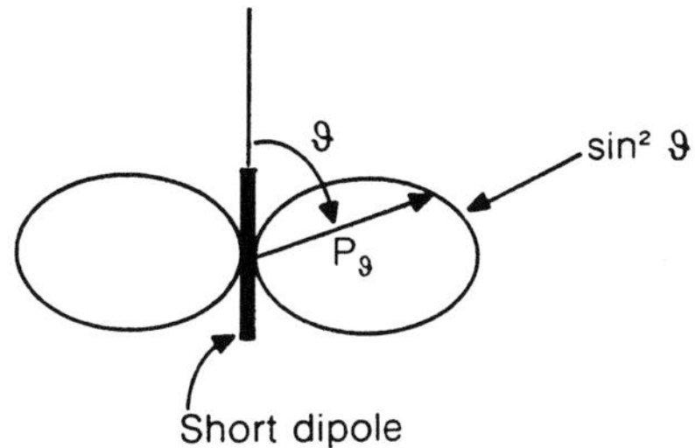

Figure 8.5 Radiation pattern for a short dipole.

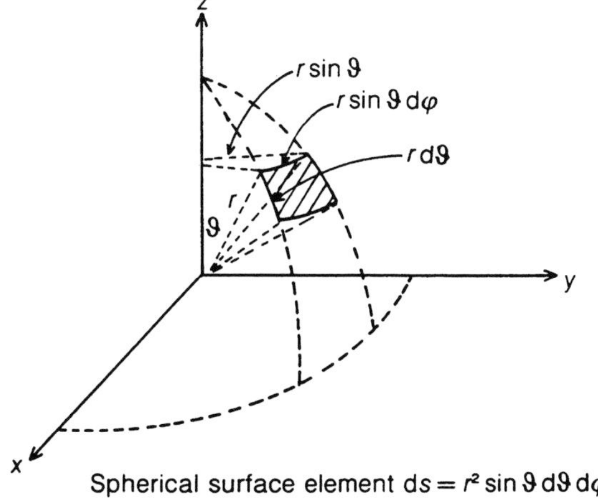

Figure 8.6 A spherical surface element.

effectively as a resistive load. This radiation resistance, R_r, is defined from the expression $(I/\sqrt{2})^2 \times R_r = P_{\text{total}}$ and, after substituting,

$$R_r = \frac{\omega^2 \mu l^2}{6\pi v_0} \qquad (\Omega).$$

If the dipole is immersed in air

$$R_r = 790(l/\lambda)^2 \qquad (\Omega).$$

Very small radiation resistances are obtained for short dipoles. This together with the reactive (capacitive in this case) component of the impedance, which has not been calculated here, makes difficult the matching of the short dipole into standard transmission lines (50 or 75 Ω characteristic impedance). Efficiency is also low because ohmic losses in the wire used to construct the dipole become comparable to its radiation resistance.

REMARKS

In understanding the performance of radiating systems, it is necessary to take full account of finite propagation velocity effects. If the current distribution on a radiating structure can be determined (say by physical intuition) the potential functions, field components and power flow can then be determined. Difficulties that may arise are those due to mathematical complexity. The antenna can then be treated as a circuit element (e.g. by calculating its radiation resistance). Care must be taken during this process to avoid confusion between lumped parameter concepts and the behaviour of radiating systems. After this process has been completed satisfactorily, the entire source–transmission line–antenna system can be studied and optimized.

In many practical cases, or when high-accuracy results are required, the current distribution on the antenna cannot be determined in advance. Examples of such cases are when the antenna is close to other objects or when the diameter of the wire used to construct the antenna is an appreciable proportion of its length. In such cases resort must be made to numerical methods (see Chapter 6).

Although the treatment of antennas in this chapter always referred to 'radiators', the ideas developed apply also to antennas used as receivers.

PROBLEMS

P66 A transmitting antenna radiates uniformly in air a power of 75 W. Calculate the total power received by a receiving antenna of an effective cross-sectional area of 5 m^2 placed at a distance of 30 km. A receiver of input impedance 50 Ω is connected to the receiving antenna and represents a matched load to it. Calculate the rms value of the voltage at the input to the receiver and the peak values of the electric and magnetic fields at the receiving antenna.
(*Answer*: 1.29 mV, 2.23 mV m^{-1}, 5.93 μA m^{-1}).

P67 A current flows on a large current sheet as shown in Figure P67. The current density is $j_y = j_0 \sin \omega t$, in Am^{-1} along the z-direction, and has constant magnitude and phase everywhere on the sheet. Examine whether waves can be launched by this structure and obtain formulae for the electric and magnetic fields and the power flux P at a distance x from the current sheet. Calculate the radiation resistance R_r of this sheet defined as $R_r = P/(j_0/\sqrt{2})^2$.
Hint: seek appropriate solutions to the wave equation for H_z. $H_z(x=0)$ may be found by applying Ampere's law on ABCDA where lengths BC and DA tend to zero.
(*Answer:* For $x > 0$ $H_z = \frac{1}{2} j_0 \sin(\omega t - kx)$, $E_y = 377\, H_z$, $R_r \simeq 188\,\Omega$, $P = E_y^2/377$.)

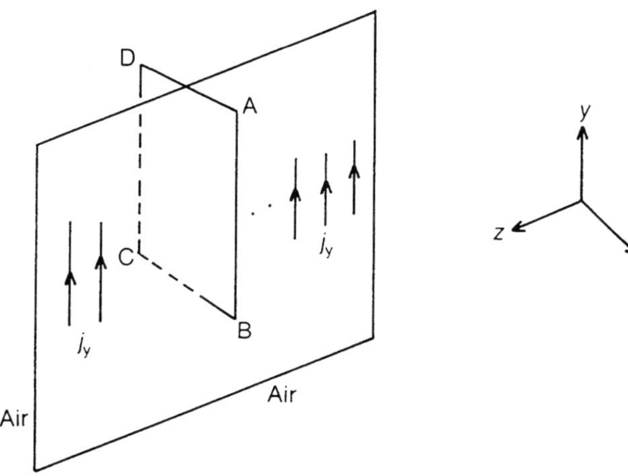

Figure P67

P68 (*a*) Show that on the equatorial plane ($\vartheta = \pi/2$) of a short dipole (Figure P68(*a*)) in air, the electric field in the far field region is

$$\bar{E}_\vartheta = \mathrm{j}\frac{60\pi Il}{r\lambda}\mathrm{e}^{\mathrm{j}(\omega t - kx)}.$$

(*b*) Two such dipoles carry opposite currents as shown in Figure P68(*b*). Show that the electric field on the equatorial plane due to the combined effect of the two dipoles is given by the formula

$$\bar{E}_\vartheta = \frac{60\pi Il}{r\lambda} 2\sin\left(\frac{ks}{2}\right)\mathrm{e}^{\mathrm{j}(\omega t - kr)}.$$

(*c*) For the configuration described in (*b*) calculate the magnitude of E_ϑ when $s = \lambda/2$ and $s = \lambda$.
Hint: Use the formula in (*a*) and superposition to work out the answer to (*b*). Assume that as far as amplitude variations are concerned $r_1 \sim r_2 \sim r$, but for phase variations $r_1 = r + s/2$, $r_2 = r - s/2$.
(*Answer:* (*c*) for $s = \lambda/2$ $E_\vartheta =$ maximum; for $s = \lambda$, $E_\vartheta = 0$.)

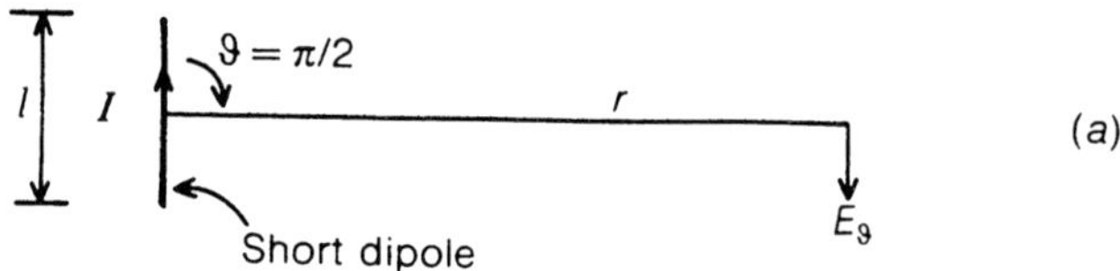

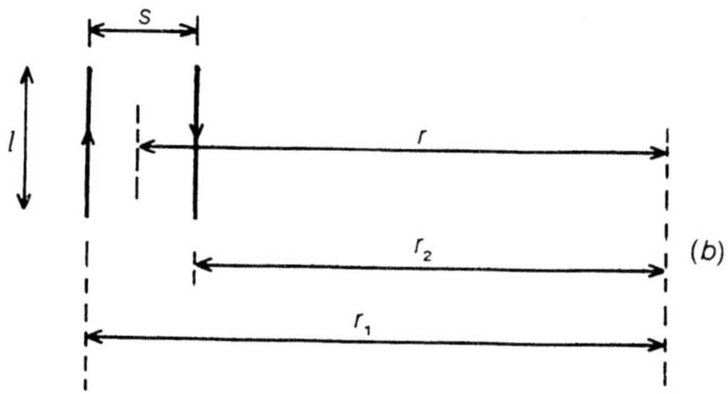

Figure P68 (a) A short dipole and (b) an array of two short dipoles.

P69 (a) Show that the formula given in problem P68 (b) reduces to

$$|E_\vartheta| = \frac{120\pi^2 Ils}{r\lambda^2}$$

when the spacing is much smaller than the wavelength ($s \ll \lambda$).
(b) Two parallel tracks on a printed circuit board are spaced at $s = 3$ mm and their length is $l = 100$ mm. They carry opposite currents of peak magnitude of 100 mA and frequency $f = 30$ MHz. Using the result in (a) estimate the strength of the electric field 10 m away. Assume that far-field conditions have been established at this distance.
(*Answer:* 35.5 μV m^{-1}.)

P70 An elementary antenna is placed above an infinite conducting plane as shown

Figure P70 Influence of ground on the radiation pattern of a dipole.

in Figure P70(a, b). For the two configurations shown study qualitatively how the radiation pattern of the antenna will be affected by the presence of the conducting plane.
Hint: use images to obtain an equivalent configuration of antennas (original plus image), where the conducting plane is absent. Explore the structure of the radiation pattern from such configurations.

APPENDIX A

A Proof of Gauss's Law

Before attempting to prove Gauss's law, it is necessary to explain the concept of a solid angle.

Consider two intersecting lines oa and ob, as shown in Figure A1(*a*). Two arcs are drawn with radii r_1, r_2 as shown. It turns out that the ratio of arc length to radius $\alpha_1\beta_2/r_1 = \alpha_2\beta_2/r_2$, is independent of the radius of the arc. This quantity (arc length/radius) is called the angle between lines Oa and Ob, and it is measured in radians.

Consider now the conical surface AOB, as shown in Figure A1(*b*), and draw a spherical surface of radius r_1. The intersection of the cone with the spherical surface defines an area ΔS_1 on the spherical surface as shown. Similarly an area ΔS_2 is defined on the spherical surface of radius r_2. It turns out that the quantity $\Delta S_1/r_1^2$ is equal to $\Delta S_2/r_2^2$, independent of the radius of the sphere. This quantity (area/radius squared) defines the solid angle with vertex at O. It is measured in steradians, the maximum angle being 4π steradians. It is also reasonable to expect that if in Figure A1(*a*) the angle is small, then, length of line segment $\alpha_1\beta_1 \cong$ arc length $\alpha_1\beta_1 = \mathrm{r}_1 \times$ (angle). Similarly, from Figure A1(*b*) for a small solid angle, the area ΔS_1 measured on the spherical surface will be approximately equal to the area defined by the cone and a plane tangent to the spherical surface at r_1.

Let us now get back to Gauss's law. Consider a point charge Q as shown in Figure A2. The surface S is arbitrary but it is chosen to completely enclose Q. A small

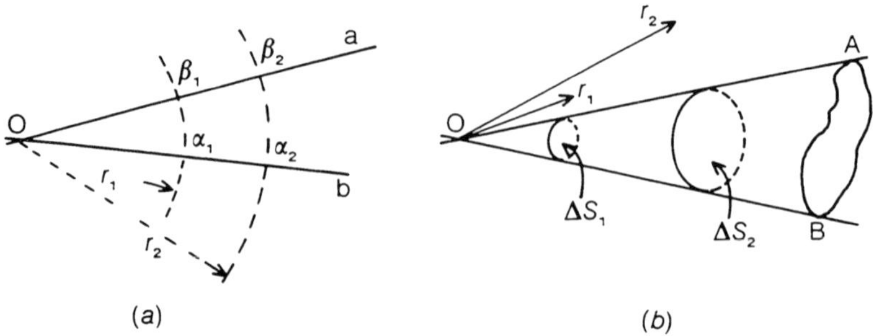

Figure A1 (*a*) Angle and (*b*) solid angle.

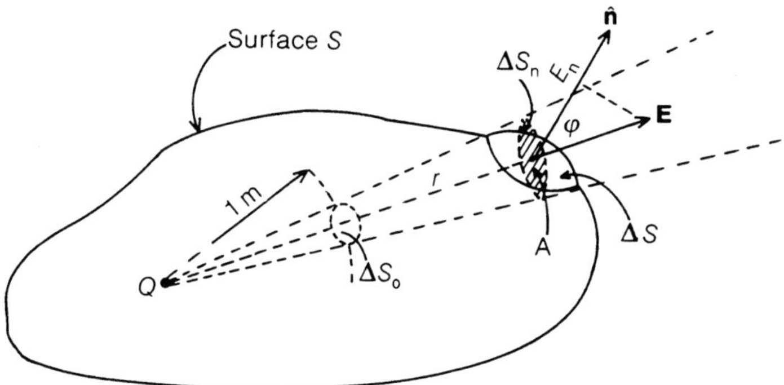

Figure A2 Construction used to prove Gauss's law.

patch of area ΔS is shown around point A on surface S. The direction normal to ΔS is $\hat{\mathbf{n}}$ and in general it may not coincide with the direction of $\mathbf{E}$. Let us call φ the angle between $\mathbf{E}$ and $\hat{\mathbf{n}}$. The projection of the patch ΔS on a plane passing from point A and perpendicular to $\mathbf{E}$ is ΔS_n (shown hatched in Figure A2) and its area is $\Delta S_n = \Delta S \cos\varphi$. Similarly the projection of $\mathbf{E}$ in the direction of $\hat{\mathbf{n}}$ is $E_n = E\cos\varphi$. It therefore follows that

$$\varepsilon_0 E_n \Delta S = \varepsilon_0 (E\cos\varphi)\Delta S = \varepsilon_0 E \Delta S_n = \varepsilon_0 \frac{Q}{4\pi\varepsilon_0 r^2}\Delta S_n = \frac{Q}{4\pi}\frac{\Delta S_n}{r^2}.$$

The solid angle defined by the position of the charge Q and ΔS_n defines also an area ΔS_0 at a distance 1 m away from Q (Figure A2). As already explained, for small angles, solid angle $= \Delta S_0/1^2 = \Delta S_n/r^2$, hence substituting

$$\varepsilon_0 E_n \Delta S = \frac{Q}{4\pi}\Delta S_0.$$

Summation of these terms for all surface patches covering S gives

$$\sum_{\substack{\text{over}\\S}} \varepsilon_0 E_n \Delta S = \frac{Q}{4\pi}\sum_{\substack{\text{over}\\S}} \Delta S_0$$

$$= \frac{Q}{4\pi} \times \text{(surface area of sphere of unit radius)}$$

$$= \frac{Q}{4\pi} \times 4\pi = Q.$$

This proves Gauss's law. If more than one charge is present, the validity of the law can be established by superposition.

APPENDIX B

Useful Mathematical Formulae

Vector Operations
Cartesian coordinates (x, y, z)

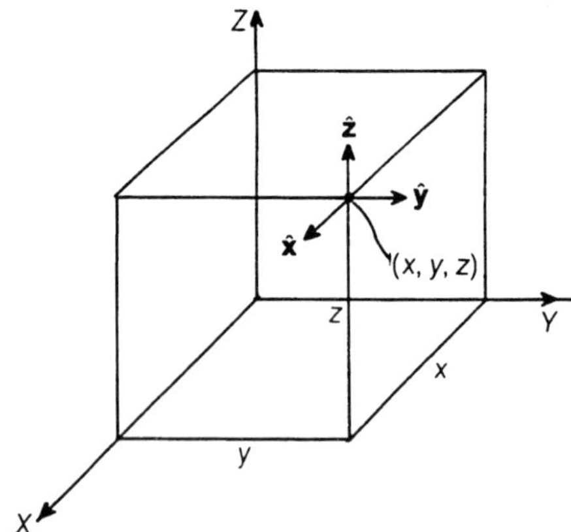

$$\Phi = \Phi(x, y, z)$$

$$\mathbf{G} = \hat{\mathbf{x}}G_x(x, y, z) + \hat{y}(G_y(x, y, z) + \hat{\mathbf{z}}G_z(x, y, z)$$
$$= \hat{\mathbf{x}}G_x + \hat{\mathbf{y}}G_y + \hat{\mathbf{z}}G_z$$

$$\nabla\Phi = \hat{\mathbf{x}}\frac{\partial\Phi}{\partial x} + \hat{\mathbf{y}}\frac{\partial\Phi}{\partial y} + \hat{\mathbf{z}}\frac{\partial\Phi}{\partial z}$$

$$\nabla\cdot\mathbf{G} = \frac{\partial G_x}{\partial x} + \frac{\partial G_y}{\partial y} + \frac{\partial G_z}{\partial z}$$

$$\nabla\times\mathbf{G} = \hat{\mathbf{x}}\left(\frac{\partial G_z}{\partial y} - \frac{\partial G_y}{\partial z}\right) + \hat{\mathbf{y}}\left(\frac{\partial G_x}{\partial z} - \frac{\partial G_z}{\partial x}\right) + \hat{\mathbf{z}}\left(\frac{\partial G_y}{\partial x} - \frac{\partial G_x}{\partial y}\right)$$

$$\nabla^2\Phi = \frac{\partial^2\Phi}{\partial x^2} + \frac{\partial^2\Phi}{\partial y^2} + \frac{\partial^2\Phi}{\partial z^2}$$

$$\nabla^2\mathbf{G} = \hat{\mathbf{x}}\nabla^2 G_x + \hat{\mathbf{y}}\nabla^2 G_y + \hat{\mathbf{z}}\nabla^2 G_z$$

Cylindrical coordinates (r, φ, z)

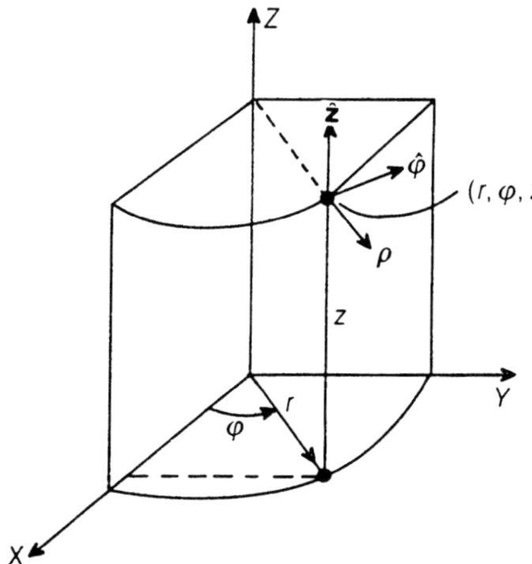

$$\Phi = \Phi(r, \varphi, z)$$

$$\mathbf{G} = \hat{\mathbf{r}}G_r + \hat{\boldsymbol{\varphi}}G_\varphi + \hat{\mathbf{z}}G_z$$

Volume element $\mathrm{d}v = r\,\mathrm{d}\varphi\,\mathrm{d}r\,\mathrm{d}z$

$$\nabla\Phi = \hat{\mathbf{r}}\frac{\partial\Phi}{\partial r} + \hat{\boldsymbol{\varphi}}\frac{1}{r}\frac{\partial\Phi}{\partial\varphi} + \hat{\mathbf{z}}\frac{\partial\Phi}{\partial z} \qquad \nabla\cdot\mathbf{G} = \frac{1}{r}\frac{\partial}{\partial r}(rG_r) + \frac{1}{r}\frac{\partial G_\varphi}{\partial\varphi} + \frac{\partial G_z}{\partial z}$$

$$\nabla\times\mathbf{G} = \hat{\mathbf{r}}\left(\frac{1}{r}\frac{\partial G_z}{\partial\varphi} - \frac{\partial G_\varphi}{\partial z}\right) + \hat{\boldsymbol{\varphi}}\left(\frac{\partial G_r}{\partial z} - \frac{\partial G_z}{\partial r}\right) + \hat{\mathbf{z}}\left(\frac{1}{r}\frac{\partial(rG_\varphi)}{\partial r} - \frac{1}{r}\frac{\partial G_r}{\partial\varphi}\right)$$

$$\nabla^2\Phi = \frac{1}{r}\frac{\partial}{\partial r}\left(r\frac{\partial\Phi}{\partial r}\right) + \frac{1}{r^2}\frac{\partial^2\Phi}{\partial\varphi^2} + \frac{\partial^2\Phi}{\partial z^2}$$

$$\nabla^2\mathbf{G} = \hat{\mathbf{r}}\left(\nabla^2G_r - \frac{2}{r^2}\frac{\partial G_\varphi}{\partial\varphi} - \frac{G_r}{r^2}\right) + \hat{\boldsymbol{\varphi}}\left(\nabla^2G_\varphi + \frac{2}{r^2}\frac{\partial G_r}{\partial\varphi} - \frac{G_\varphi}{r^2}\right) + \hat{\mathbf{z}}(\nabla^2G_z)$$

Spherical coordinates (r, ϑ, φ)

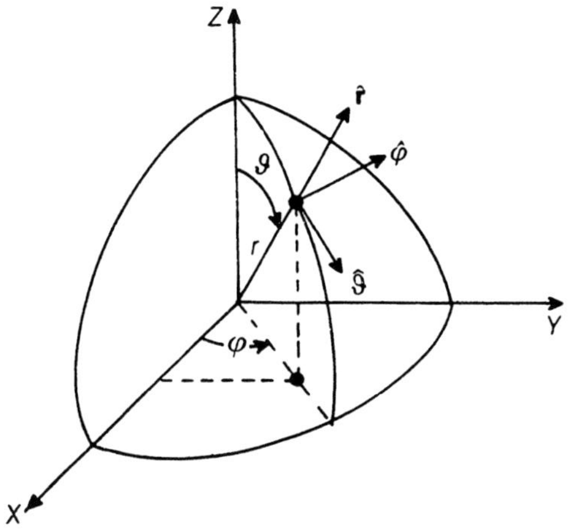

$$\Phi = \Phi(r, \vartheta, \varphi)$$

$$\mathbf{G} = \hat{\mathbf{r}}G_r + \hat{\boldsymbol{\vartheta}}G_\vartheta + \hat{\boldsymbol{\varphi}}G_\varphi$$

Volume element $\mathrm{d}v = r^2\sin\theta\mathrm{d}r\,\mathrm{d}\vartheta\,\mathrm{d}\varphi$

$$\nabla\Phi = \hat{\mathbf{r}}\frac{\partial\Phi}{\partial r} + \hat{\boldsymbol{\vartheta}}\frac{1}{r}\frac{\partial\Phi}{\partial\vartheta} + \frac{\hat{\boldsymbol{\varphi}}}{r\sin\vartheta}\frac{\partial\Phi}{\partial\varphi}$$

$$\nabla\cdot\mathbf{G} = \frac{1}{r^2}\frac{\partial}{\partial r}(r^2 G_r) + \frac{1}{r\sin\vartheta}\frac{\partial}{\partial\vartheta}(\sin\vartheta\, G_\vartheta) + \frac{1}{r\sin\vartheta}\frac{\partial G_\varphi}{\partial\varphi}$$

$$\nabla\times\mathbf{G} = \frac{\hat{\mathbf{r}}}{r\sin\vartheta}\left(\frac{\partial}{\partial\vartheta}(G_\vartheta\sin\vartheta) - \frac{\partial G_\vartheta}{\partial\varphi}\right) + \frac{\hat{\boldsymbol{\vartheta}}}{r}\left(\frac{1}{\sin\vartheta}\frac{\partial G_r}{\partial\varphi} - \frac{\partial}{\partial r}(rG_\varphi)\right) + \frac{\hat{\boldsymbol{\varphi}}}{r}\left(\frac{\partial}{\partial r}(rG_\vartheta) - \frac{\partial G_r}{\partial\vartheta}\right)$$

$$\nabla^2\Phi = \frac{1}{r^2}\frac{\partial}{\partial r}\left(r^2\frac{\partial\Phi}{\partial r}\right) + \frac{1}{r^2\sin\vartheta}\frac{\partial}{\partial\vartheta}\left(\sin\vartheta\frac{\partial\Phi}{\partial\vartheta}\right) + \frac{1}{r^2\sin^2\vartheta}\frac{\partial^2\Phi}{\partial\varphi^2}$$

$$\nabla^2\mathbf{G} = \hat{\mathbf{r}}\left[\nabla^2 G_r - \frac{2}{r^2}\left(G_r + \cot\vartheta\, G_\vartheta + \operatorname{cosec}\vartheta\frac{\partial G_\varphi}{\partial\varphi} + \frac{\partial G_\vartheta}{\partial\vartheta}\right)\right] + \hat{\boldsymbol{\vartheta}}\left[\nabla^2 G_\vartheta - \frac{1}{r^2}\left(\operatorname{cosec}^2\vartheta\, G_\vartheta - 2\frac{\partial G_r}{\partial\vartheta} + 2\cot\vartheta\operatorname{cosec}\vartheta\frac{\partial G_\vartheta}{\partial\varphi}\right)\right] + \hat{\boldsymbol{\varphi}}\left[\nabla^2 G_\varphi - \frac{1}{r^2}\left(\operatorname{cosec}^2\vartheta\, G_\varphi - 2\operatorname{cosec}\vartheta\frac{\partial G_r}{\partial\varphi} - 2\cot\vartheta\operatorname{cosec}\vartheta\frac{\partial G_\vartheta}{\partial\varphi}\right)\right]$$

Some Vector Identities

$$\nabla\cdot\nabla\times\mathbf{G} = 0$$

$$\nabla\times\nabla\Phi = 0$$

$$\nabla\times\nabla\times\mathbf{G} = \nabla(\nabla\cdot\mathbf{G}) - \nabla^2\mathbf{G}$$

$$\nabla\cdot(\Phi\mathbf{G}) = \mathbf{G}\cdot\nabla\Phi + \Phi\nabla\cdot\mathbf{G}$$

$$\nabla\times(\Phi\mathbf{G}) = \nabla\Phi\times\mathbf{G} + \Phi\nabla\times\mathbf{G}$$

$$\nabla\cdot\nabla\Phi = \nabla^2\Phi$$

$$\mathbf{A}\times(\mathbf{B}\times\mathbf{C}) = \mathbf{B}(\mathbf{A}\cdot\mathbf{C}) - \mathbf{C}(\mathbf{A}\cdot\mathbf{B})$$

$$\mathbf{A}\cdot(\mathbf{B}\times\mathbf{C}) = \mathbf{B}\cdot(\mathbf{C}\times\mathbf{A}) = \mathbf{C}\cdot(\mathbf{A}\times\mathbf{B})$$

Some Basic Derivatives and Integrals

$$\frac{\mathrm{d}}{\mathrm{d}x}(x^n) = nx^{n-1}$$

$$\frac{\mathrm{d}}{\mathrm{d}x}\left(\frac{f(x)}{g(x)}\right) = \frac{g(x)f'(x) - g'(x)f(x)}{g^2(x)}$$

$$\frac{\mathrm{d}}{\mathrm{d}x}[\ln g(x)] = \frac{g'(x)}{g(x)}$$

$$\frac{\mathrm{d}}{\mathrm{d}x}(\sin \alpha x) = \alpha \cos \alpha x \qquad \frac{\mathrm{d}}{\mathrm{d}x}(\cos \alpha x) = -\alpha \sin \alpha x$$

$$\frac{\mathrm{d}}{\mathrm{d}x}[f(u)] = \frac{\mathrm{d}f}{\mathrm{d}u}\frac{\mathrm{d}u}{\mathrm{d}x}$$

$$\int x^{\alpha}\mathrm{d}x = \frac{x^{\alpha+1}}{\alpha+1} \qquad \int \frac{\mathrm{d}x}{x} = \ln x$$

$$\int \cos x \,\mathrm{d}x = \sin x \qquad \int \sin x \,\mathrm{d}x = -\cos x$$

$$\int \frac{\mathrm{d}x}{ax^2+b} = \frac{1}{2\sqrt{-ab}} \ln \frac{x\sqrt{a}-\sqrt{-b}}{x\sqrt{a}+\sqrt{-b}} \qquad \text{for} \quad a>0,\ b<0$$

$$\int \frac{\mathrm{d}x}{ax+b} = \frac{\ln(ax+b)}{a}$$

APPENDIX C

Useful Physical Constants and Other Data

Constants

Permittivity of free spare $\varepsilon_0 = 8.8542 \times 10^{-12}\,\mathrm{F\,m^{-1}}$
Permeability of free space $\mu_0 = 4\pi \times 10^{-7}\,\mathrm{H\,m^{-1}}$
Magnitude of electronic charge $e = 1.6 \times 10^{-19}\,\mathrm{C}$
Electron mass $M_0 = 9.1 \times 10^{-31}\,\mathrm{kg}$
Speed of light in vacuum $c = 3 \times 10^{8}\,\mathrm{m\,s^{-1}}$

Typical values

Insulating liquids $\varepsilon_r \simeq 3$
Insulating solids $\varepsilon_r \simeq 5$
Mild steel $\mu_r \simeq 500$
Electrical conductivity of copper $\sigma = 5.8 \times 10^{7}\,\mathrm{S\,m^{-1}}$
Breakdown strength of air under standard conditions $E_{max} = 30\,\mathrm{kV\,cm^{-1}}$
Saturation magnetic flux density of commercial steel $B_{max} \simeq 1.2\,\mathrm{T}$

List of commonly used units in the *SI* system

Quantity	Unit name	Unit symbol
Length	metre	m
Mass	kilogram	kg
Time	second	s
Electric current	ampere	A
Force	newton	N
Energy	joule	J
Power	watt	W

Quantity	Unit name	Unit symbol
Electric potential	volt	V
Electric charge	coulomb	C
Magnetic flux	weber	Wb
Magnetic flux density	tesla	T
Resistance	ohm	Ω
Inductance	henry	H
Capacitance	farad	F
Electric field strengh	volt per metre	$\mathrm{V\,m^{-1}}$
Magnetic field strengh	ampere per metre	$\mathrm{A\,m^{-1}}$

APPENDIX D

Further Reading

It is impossible to give a complete list of books consulted by the author. The following may be found useful by the reader looking for more detail or an alternative approach.

INTRODUCTORY TEXTS

Electromagnetism for Engineers—An Introductory Course
by P. Hammond, Pergammon Press.
Electricity and Magnetism and *Electromagnetic Waves*
both by E.R. Dobbs, Routledge and Kegan Paul.

ADVANCED TEXTS

Electromagnetic Field Theory—A Problem Solving Approach
by M. Zahn, John Wiley
Advanced Engineering Electromagnetics
by C.A. Balanis, John Wiley
Electromagnetic Fields and Energy
by H.A. Haus and J.R. Melcher, Prentice Hall
Fields and Waves in Communication Electronics
by S. Ramo, J.R. Whinnery and T. van Duzer, John Wiley
Antenna Theory—Analysis and Design
by C.A. Balanis, Harper and Row

Index